Reviews of Environmental Contamination and Toxicology

VOLUME 149

Springer
New York
Berlin
Heidelberg
Barcelona
Budapest
Hong Kong
London
Milan
Paris
Santa Clara
Singapore
Tokyo

Reviews of Environmental Contamination and Toxicology

Continuation of Residue Reviews

Editor
George W. Ware

Founding Editor
Francis A. Gunther

VOLUME 149

Springer

Coordinating Board of Editors

Springer-Verlag
New York: 175 Fifth Avenue, New York, NY 10010, USA
Heidelberg: 69042 Heidelberg, Postfach 10 52 80, Germany

Library of Congress Catalog Card Number 62-18595.
Printed in the United States of America.

ISSN 0179-5953

Printed on acid-free paper.

ISBN 0-387-94863-5 Springer-Verlag New York Berlin Heidelberg   SPIN 10523385

Foreword

International concern in scientific, industrial, and governmental communities over traces of xenobiotics in foods and in both abiotic and biotic environments has justified the present triumvirate of specialized publications in this field: comprehensive reviews, rapidly published research papers and progress reports, and archival documentations. These three international publications are integrated and scheduled to provide the coherency essential for nonduplicative and current progress in a field as dynamic and complex as environmental contamination and toxicology. This series is reserved exclusively for the diversified literature on "toxic" chemicals in our food, our feeds, our homes, recreational and working surroundings, our domestic animals, our wildlife and ourselves. Tremendous efforts worldwide have been mobilized to evaluate the nature, presence, magnitude, fate, and toxicology of the chemicals loosed upon the earth. Among the sequelae of this broad new emphasis is an undeniable need for an articulated set of authoritative publications, where one can find the latest important world literature produced by these emerging areas of science together with documentation of pertinent ancillary legislation.

Research directors and legislative or administrative advisers do not have the time to scan the escalating number of technical publications that may contain articles important to current responsibility. Rather, these individuals need the background provided by detailed reviews and the assurance that the latest information is made available to them, all with minimal literature searching. Similarly, the scientist assigned or attracted to a new problem is required to glean all literature pertinent to the task, to publish new developments or important new experimental details quickly, to inform others of findings that might alter their own efforts, and eventually to publish all his/her supporting data and conclusions for archival purposes.

In the fields of environmental contamination and toxicology, the sum of these concerns and responsibilities is decisively addressed by the uniform, encompassing, and timely publication format of the Springer-Verlag (Heidelberg and New York) triumvirate:

Reviews of Environmental Contamination and Toxicology [Vol. 1 through 97 (1962–1986) as Residue Reviews] for detailed review articles concerned with any aspects of chemical contaminants, including pesticides, in the total environment with toxicological considerations and consequences.

Bulletin of Environmental Contamination and Toxicology (Vol. 1 in 1966)
for rapid publication of short reports of significant advances and discoveries in the fields of air, soil, water, and food contamination and pollution as well as methodology and other disciplines concerned with the introduction, presence, and effects of toxicants in the total environment.

Archives of Environmental Contamination and Toxicology (Vol. 1 in 1973)
for important complete articles emphasizing and describing original experimental or theoretical research work pertaining to the scientific aspects of chemical contaminants in the environment.

Manuscripts for *Reviews* and the *Archives* are in identical formats and are peer reviewed by scientists in the field for adequacy and value; manuscripts for the *Bulletin* are also reviewed, but are published by photo-offset from camera-ready copy to provide the latest results with minimum delay. The individual editors of these three publications comprise the joint Coordinating Board of Editors with referral within the Board of manuscripts submitted to one publication but deemed by major emphasis or length more suitable for one of the others.

                                        Coordinating Board of Editors

Preface

Thanks to our news media, today's lay person may be familiar with such environmental topics as ozone depletion, global warming, greenhouse effect, nuclear and toxic waste disposal, massive marine oil spills, acid rain resulting from atmospheric SO_2 and NO_x, contamination of the marine commons, deforestation, radioactive leaks from nuclear power generators, free chlorine and CFC (chlorofluorocarbon) effects on the ozone layer, mad cow disease, pesticide residues in foods, green chemistry or green technology, volatile organic compounds (VOCs), hormone- or endocrine-disrupting chemicals, declining sperm counts, and immune system suppression by pesticides, just to cite a few. Some of the more current, and perhaps less familiar, additions include *xenobiotic transport, solute transport, Tiers 1 and 2, USEPA to cabinet status, and zero-discharge.* These are only the most prevalent topics of national interest. In more localized settings, residents are faced with leaking underground fuel tanks, movement of nitrates and industrial solvents into groundwater, air pollution and "stay-indoors" alerts in our major cities, radon seepage into homes, poor indoor air quality, chemical spills from overturned railroad tank cars, suspected health effects from living near high-voltage transmission lines, and food contamination by "flesh-eating" bacteria and other fungal or bacterial toxins.

It should then come as no surprise that the '90s generation is the first of mankind to have become afflicted with *chemophobia*, the pervasive and acute fear of chemicals.

There is abundant evidence, however, that virtually all organic chemicals are degraded or dissipated in our not-so-fragile environment, despite efforts by environmental ethicists and the media to persuade us otherwise. However, for most scientists involved in environmental contaminant reduction, there is indeed room for improvement in all spheres.

Environmentalism is the newest global political force, resulting in the emergence of multi-national consortia to control pollution and the evolution of the environmental ethic. Will the new politics of the 21st century be a consortium of technologists and environmentalists or a progressive confrontation? These matters are of genuine concern to governmental agencies and legislative bodies around the world, for many serious chemical incidents have resulted from accidents and improper use.

For those who make the decisions about how our planet is managed, there is an ongoing need for continual surveillance and intelligent controls to avoid endangering the environment, the public health, and wildlife. Ensuring safety-in-use of the many chemicals involved in our highly industrialized culture is a dynamic challenge, for the old, established materials are continually being displaced by newly developed molecules more acceptable to federal and state regulatory agencies, public health officials, and environmentalists.

Adequate safety-in-use evaluations of all chemicals persistent in our air, foodstuffs, and drinking water are not simple matters, and they incorporate the judgments of many individuals highly trained in a variety of complex biological, chemical, food technological, medical, pharmacological, and toxicological disciplines.

Reviews of Environmental Contamination and Toxicology continues to serve as an integrating factor both in focusing attention on those matters requiring further study and in collating for variously trained readers current knowledge in specific important areas involved with chemical contaminants in the total environment. Previous volumes of *Reviews* illustrate these objectives.

Because manuscripts are published in the order in which they are received in final form, it may seem that some important aspects of analytical chemistry, bioaccumulation, biochemistry, human and animal medicine, legislation, pharmacology, physiology, regulation, and toxicology have been neglected at times. However, these apparent omissions are recognized, and pertinent manuscripts are in preparation. The field is so very large and the interests in it are so varied that the Editor and the Editorial Board earnestly solicit authors and suggestions of underrepresented topics to make this international book series yet more useful and worthwhile.

Reviews of Environmental Contamination and Toxicology attempts to provide concise, critical reviews of timely advances, philosophy, and significant areas of accomplished or needed endeavor in the total field of xenobiotics in any segment of the environment, as well as toxicological implications. These reviews can be either general or specific, but properly they may lie in the domains of analytical chemistry and its methodology, biochemistry, human and animal medicine, legislation, pharmacology, physiology, regulation, and toxicology. Certain affairs in food technology concerned specifically with pesticide and other food-additive problems are also appropriate subjects.

Justification for the preparation of any review for this book series is that it deals with some aspect of the many real problems arising from the presence of any foreign chemical in our surroundings. Thus, manuscripts may encompass case studies from any country. Added plant or animal pest-control chemicals or their metabolites that may persist into food and animal feeds are within this scope. Food additives (substances deliberately added to foods for flavor, odor, appearance, and preservation, as well as those inadvertently added during manufacture, packing, distribution, and storage) are also considered suitable review material. Additionally, chemical contamination in any manner of air, water, soil, or plant or animal life is within these objectives and their purview.

Normally, manuscripts are contributed by invitation, but suggested topics are welcome. Preliminary communication with the Editor is recommended before volunteered review manuscripts are submitted.

Department of Entomology   G.W.W.
University of Arizona
Tucson, Arizona

Table of Contents

Toxicology of Mono-, Di-, and Triethanolamine

J.B. Knaak*,**, Hon-Wing Leung†,**, W.T. Stott‡,**
J. Busch**, and J. Bilsky**

Contents

*Occidental Chemical Corp., 360 Rainbow Boulevard South, Niagara Falls, NY 14302, U.S.A.

†Union Carbide Corp., 39 Old Ridgebury Road, Danbury, CT 06817, U.S.A.

‡Toxicology Research Laboratory, The Dow Chemical Company, Midland, MI 48640, U.S.A.

**Chemical Manufacturers Association, 1300 Wilson Boulevard, Arlington, VA 22209, U.S.A.

© 1997 by Springer-Verlag New York, Inc.
Reviews of Environmental Contamination and Toxicology, Vol. 149.

I. Introduction

The family of ethanolamines, including monoethanolamine (MEA), diethanolamine (DEA), and triethanolamine (TEA), offers a broad spectrum of
application opportunities because they combine the properties of amines
and alcohols. Ethanolamines exhibit the unique capability of undergoing
reactions common to both groups. As amines, they are mildly alkaline and
react with acids to form salts or soaps. As alcohols, they are hygroscopic
and can be esterified. Thus, since their introduction in the late 1920s, the
ethanolamines have found uses in such diverse areas as gas sweetening,
where they remove carbon dioxide and hydrogen sulfide; metalworking, in
which they act as corrosion inhibitors in synthetic fluids; textile processing,
where they serve as intermediates for cationic softening agents, durable
press fabric finishes, dye leveling agents, lubricants, and scouring agents;
detergent and specialty cleaner formulations, in which they are used to
form various amine salts and to control pH; and a slew of other applications ranging from concrete admixtures to flexible urethane foam catalysts
to pharmaceuticals to agricultural chemicals and photographic emulsions
(Dow Chemical 1988; Howe-Grant 1992; OCC 1995).

Since the ethanolamines receive widespread applications in various in-
dustries and consumer products, they may present a significant potential
for human exposure. A considerable number of experimental studies have
been conducted over the years to understand the potential hazards of the
ethanolamines. These studies evaluate the toxicity of the ethanolamines
from single and repeated exposures, including their potential to cause mutation, birth defects, and tumors. These extensive toxicology data have been
exhaustively compiled, fully annotated, and critically reviewed in this comprehensive paper.

II. Physical and Chemical Properties

The physical and chemical properties of the ethanolamines are summarized
in Table 1. Pure ethanolamines are colorless liquids possessing a mild am-
moniacal odor. Their chemical and physical properties exhibit properties of
their two functional groups, alcohol and amines. Figure 1 gives their structures. The ethanolamines are soluble in most polar solvents (e.g., water,
alcohol, etc.) but have limited solubility in organic solvents such as n-
heptane (<0.06 g/100 g). Triethanolamine is the most polar, with its
three –OH groups, followed by DEA and MEA. All three solvents are
viscous at low temperatures, with MEA being considerably less viscous than

Table 1. Physical and chemical properties of the alkanolamines.

Property	Mono- ethanolamine	Diethanolamine	Triethanolamine
CAS Number	141-43-5	111-42-2	102-71-6
Chemical name	Ethanol, 2-amino	Ethanol, 2,2- iminobis	Ethanol, 2,2,2 ni- trilo tris
Molecular weight	61.08	105.13	149.18
Physical state (25°C)	Liquid above 10.5°C	Liquid above 28°C	Liquid above 20.5°C
Melting point (°C)	10.5	28.0	20.5
Boiling point (°C)	171	268.8	335.4
Density (g/cm^3)	1.0180 at 20°C	1.0966 at 20°C	1.1242 at 20°C
Refractive index	1.4542 at 20°C	1.4776 at 20°C	1.4853 at 20°C
Solubility	Water, alcohol, ethanol, ben- zene, chloro- form, glycerol, ligroin	Water, alcohol, ethanol, ben- zene	Water, alcohol, ethanol, ben- zene, chloro- form, ligroin
pK_a	9.5 at 25°C	8.88 at 25°C	7.76 at 25°C
Mass spectrum, NIST	34160	34166	4612
m/e dominant species	61.05	105.08	149.11
IR reference	ALIRS 5776	COB 5638	COB 6371

CAS, Chemical Abstracts Service; NIST, U. S. National Institute of Standards and Technology; ALIRS, Aldrich Library of Infrared Spectra, Ed. III. Aldrich Chemical Co., Milwaukee, WI; COB, COBLENTZ Collection, Joint Committee on Atomic and Molecular Physical Data; Evaluated IR Spectra, BioRad Laboratories, Stadtler Division, Philadelphia, PA.

Properties from Lange's Handbook of Chemistry, 14th ed. McGraw-Hill Inc., New York, 1992.

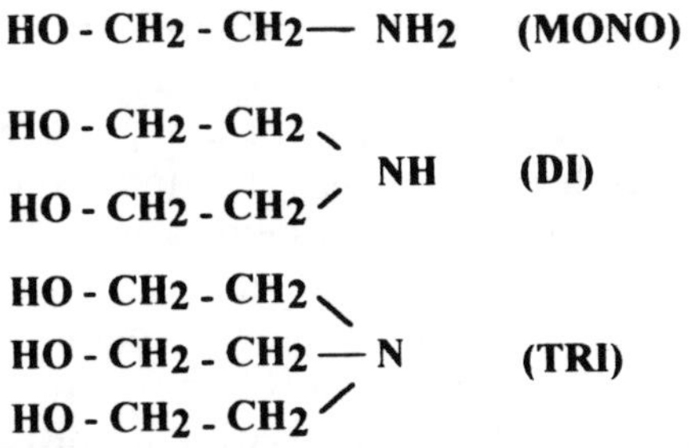

HO - CH$_2$ - CH$_2$— NH$_2$ (MONO)

HO - CH$_2$ - CH$_2$ ＼
                              NH   (DI)
HO - CH$_2$ - CH$_2$ ／

HO - CH$_2$ - CH$_2$ ＼
HO - CH$_2$ - CH$_2$ — N   (TRI)
HO - CH$_2$ - CH$_2$ ／

Fig. 1. Structure of mono-, di-, and triethanolamines.

DEA or TEA. Their melting points range from 10.5° to 28 °C, with MEA having the lowest melting point, DEA the highest, and TEA falling in between at 20.5 °C; DEA is a white crystalline solid at room temperature.

All three alkanolamines are alkaline, having pH values ranging from 10.0 to 12.5. MEA is the most alkaline and is similar in strength to aqueous ammonia. The ethanolamines boil at elevated temperatures ranging from 171° to 335.4 °C, with MEA having the lowest and TEA the highest boiling point. Although ethanolamines have relatively high freezing points, they are warmed to decrease their viscosity during transport and handling. The alkanolamines are unstable in the presence of reactive metals and water at elevated temperatures, reacting to produce colored products, and under extreme conditions react to liberate hydrogen. Because of their instability at higher temperatures, flammability limits are hard to determine. Their flash point lies between 190 °F for MEA and 400 °F for TEA. APHA (American Public Health Association) color is determined using an ASTM (American Society for Testing Materials) D1209 cobalt–platinum method. MEA is generally water-white to off-white in color. DEA has a specification limit of 50 APHA, or slightly yellow. TEA can range from off-white to dark amber in color (250 APHA) depending on the manner in which it was produced and stored. Detailed analytical procedures are available in the literature (Dow Chemical 1988) or from individual manufacturers.

III. Production, Use, and Chemistry
A. Production and Use

MEA is produced by reacting 1 mole of ethylene oxide (EO) with 1 mole of ammonia (NH_3). The addition of 2 and 3 moles of EO to 1 mole of ammonia will produce DEA and TEA, while additional EO will continue to react to produce higher EO adducts of TEA. In a typical production facility, EO is reacted with ammonia in a batch process to produce a crude mixture of roughly one-third each MEA, DEA, and TEA. The crude is stored and then separated by distillation (Dow Chemical 1988; OCC 1995). Because they have similar properties, the alkanolamines are used interchangeably or in combination with each other. Stainless steel, 316 L and 304 L, is the preferred material for constructing shipping and storage tanks. Table 2 lists the current uses of the ethanolamines as a group, while Table 3 gives the individual uses for each ethanolamine. The total worldwide production capacity for ethanolamines in 1992 was estimated at 300,000 metric tons. Of this demand, 145,450 tons was exported, 7727 tons was imported from foreign producers, and 45,455–54,545 tons was consumed in internal operations (OCC 1995). In 1995, the annual U.S. production capacity for ethanolamines was estimated to be 447,727 metric tons (SRI 1995).

Table 2. Major uses of the alkanolamines.

Applications	Percent of production	Action	
Exports	39		
Surfactants	13	Anionics (sulfonic acids):	Nonionics (fatty acids):
Gas processing	11	remove carbon dioxide	remove H_2S
Ethyleneamines	10		
Corrosion inhibitors	9	Salts of phosphoric acid	Salts of sulfuric acid:
Miscellaneous	7	Plasticizers	Corrosion inhibitors
Concrete/cement	6	Grinding aid Air entrainer	Drying accelerator
Textiles	5	Anionic fiber treatment: Acid acceptor Alkaline additives	UV light fade inhibitor: Anti-static agents Humectants/lubricants
Detergents	4	Heavy duty (industrial applications)	Light duty (dishwashing)
Agricultural chemicals	1	Herbicides	
Personal care products	1		

From OCC (1995).

B. Chemistry

1. Soaps, Surfactants, and Salts. Ethanolamines function as important intermediates in the production of surfactants because of their dual functional groups (Davidson and Milwidsky, 1968; Howe-Grant 1992; Jungerman and Tabor 1967; OCC 1995). At elevated temperatures, MEA and DEA react in a 2 : 1 ratio with long-chain fatty acids to produce Kritchevsky-type or regular ethanolamides. Regular ethanolamides consist of a viscous slurry of 60%–70% ethanolamide and lesser yields of unreacted ethanolamine and amide and amine esters according to Eq. (1).

$$C_{17}H_{35}C(O)OH + 2 H_2NCH_2CH_2OH \rightarrow C_{17}H_{35}C(O)NHCH_2CH_2OH + H_2NCH_2CH_2OH$$

Stearic Acid   MEA   (1)

$$+ C_{17}H_{35}C(O)OCH_2CH_2NH_2 + C_{17}H_{35}C(O)OCH_2CH_2NHC(O)C_{17}H_{35}$$

Table 3. Major uses of mono-, di-, and triethanolamines.

Applications	Percent of production
Monoethanolamine market	
Ethyleneamines	40
Gas purification	25
Surfactants	12
Detergents	10
Miscellaneous	7
Metalworking fluids	2
Cosmetics	2
Textile processing	2
Diethanolamine market	
Surfactants	39
Gas purification	30
Textile processing	15
Metalworking fluids	10
Miscellaneous	8
Laundry detergents	2
Agricultural chemicals	2
Triethanolamine market	
Metalworking fluids	33
Concrete/cement	25
Surfactants	20
Textile processing	8
Miscellaneous	6
Agricultural chemicals	3
Cosmetics	2

From OCC (1995).

When DEA is used in the production of ethanolamides, amine and amide
diesters and some morpholine and piperazine derivatives are produced, as
shown in Eq. (2).

$$C_{12}H_{25}C(O)OH + HN(CH_2CH_2OH)_2 \rightarrow C_{12}H_{25}C(O)N(CH_2CH_2OH)_2 + HN(CH_2CH_2OH)_2$$

Lauric Acid   DEA

$$\text{(2)}$$

$$+ HN(CH_2CH_2OC(O)C_{12}H_{25})_2 + (C_{12}H_{25}C(O)OCH_2CH_2)_2NC(O)C_{12}H_{25} + O\langle\ \rangle NH$$

$$+ HOCH_2CH_2\,N\langle\ \rangle NCH_2CH_2OH$$

At lower temperatures, MEA and DEA react with long-chain fatty acid esters in a 1 : 1 mole ratio to produce a 90 + % pure, crystalline ethanolamide mixture called a superamide [see Eq. (3)].

$$C_{12}H_{25}C(O)OCH_3 + H_2NCH_2CH_2OH \rightarrow C_{12}H_{25}C(O)NHCH_2CH_2OH + CH_3OH$$
Lauric Acid Methyl Ester   MEA

$$(3)$$

$$C_{17}H_{35}C(O)OCH_3 + HN(CH_2CH_2OH)_2 \rightarrow C_{17}H_{35}C(O)N(CH_2CH_2OH)_2 + CH_3OH$$
Steric Acid Methyl Ester   DEA

Superamides (1 : 1 ratio of reactants) contain some of the same unreacted ethanolamines and amide and amine esters but to a lesser extent. The alcohol by-product that results from using an ester in place of a fatty acid is removed from the mixture during processing. The monoethanolamides are used for "water-in-oil" type formulations where they act as foam stabilizers, corrosion inhibitors, and rinse improvers. Their largest use is in heavy-duty, dry, powdered detergents, where their higher alkalinity and lower water solubility are needed. The diethanolamides possessing lower alkalinity and better water solubility are used in liquid laundry and dishwashing detergent formulations, cosmetics, shampoos, and hair conditioners. In these applications, they act as foam improvers, thickeners, and opacifying agents.

MEA and DEA combine with long-chain fatty acids to produce neutral carboxylates or alkanolamine soaps as shown in Eq. (4).

$$C_{10}H_{19}C(O)OH + H_2NCH_2CH_2OH \rightarrow C_{10}H_{19}C(O)O^- + H_3{}^+NCH_2CH_2OH$$
Oleic Acid   MEA

$$(4)$$

$$C_{17}H_{35}C(O)OH + HN(CH_2CH_2OH)_2 \rightarrow C_{17}H_{35}C(O)^- + H_2{}^+N(CH_2CH_2OH)_2$$
Steric Acid   DEA

The soaps are generally waxy, noncrystalline, neutral to mildly alkaline, noncorrosive to metals, and not harmful to textiles. They are used extensively in formulations as detergent emulsifiers, where they help form a homogeneous mixture between hydrocarbons (grease, fat, and waxes) and water. When a higher alkalinity and limited water solubility are desired, MEA, oleic acid, steric acid, lauric acid, and tall oil are commonly used. DEA is used in light-duty liquid detergents, where high alkalinity is not desired and high solubility is needed. In products requiring high solubility and product mildness, TEA is used. TEA cannot form amides, but they do give esters at temperatures high enough to eliminate water.

The basic alkanolamines are also used in combination with organic sulfate and sulfonates to produce synthetic detergents. Equation (5) shows the formation of several ethanolamine salts.

$$C_{12}H_{25}OSO_3H + HOCH_2CH_2N(CH_2CH_2OH)_2 \rightarrow C_{12}H_{25}OSO_3^- \; H^+N(CH_2CH_2OH)_3$$

Lauryl Hydrogen Sulfate   TEA

$$C_xH_y\!\!\left\langle\bigcirc\right\rangle\!\!SO_3H + HOCH_2CH_2N(CH_2CH_2OH)_2 \rightarrow C_xH_y\!\!\left\langle\bigcirc\right\rangle\!\!SO_3^- \; H^+N(CH_2CH_2OH)_3$$

LAS   TEA

(5)

The TEA salt of lauryl hydrogen sulfate is a common anionic surfactant used in hand soaps and hand creams. Linear alkyl benzene sulfonate salts are used as replacements for alkanolamine soaps in many textile and hard-surface cleaning applications. Sulfonate salts are more resistant to hardness ions (Ca^{2+} and Mg^{2+}) than carboxylate or sulfate salts and are generally of lower acute toxicity and less irritating to the skin and eyes than the corresponding metal salts. Due to its higher water solubility and mild akalinity, TEA is the ethanolamine most commonly used for the production of these anionic surfactants.

2. Gas and Textile Processing.   MEA and DEA are used to remove acid gases, such as carbon dioxide and hydrogen sulfide, from natural gas, oil refinery gas streams, ammonia synthesis gas, and hydrocarbons from ethylene crackers. In practice, the gas streams are passed through a fluidized bed of ethanolamines and water, where the ethanolamines react with the acid gases to form water-soluble salts (Dow Chemical 1962). See Eq. (6):

$$HOCH_2CH_2NH_2 + CO_2 + H_2O \rightarrow HOCH_2CH_2N^+H_3 + HCO_3^-$$

MEA   Carbon Dioxide

$$HOCH_2CH_2NH_2 + H_2S \rightarrow HOCH_2CH_2N^+H_3 + HS^-$$

MEA   Hydrogen Sulfide

(6)

Ethanolamine may be recovered from its salt form by heating the aqueous salt solution, dissociating the salt, and volatilizing off the acid gas. The acid gases may be condensed and collected for reuse. MEA is the preferred amine for this process. However, DEA must be used if significant amounts of carbonyl sulfide are present in the gas because an unregenerable complex is formed with MEA.

   Ethanolamines are used in the textile industry in a number of applications, which include anionic fiber treatment, UV (ultraviolet) light fade inhibitor, acid acceptor, antistatic agents (Howe-Grant 1992; Polish Patent 1987; USSR Patent 1989) and miscellaneous applications. In anionic fiber treatment, the ethanolamines are used to make ethanolamides that penetrate the anionic fiber to improve the material's tear strength and resistance to abrasion. As UV light fade inhibitors, DEA and TEA are absorbed onto dyed fiber and dried; the functional groups prevent UV color fade. TEA is added to tetrakis hydroxy phosphonium chloride as an acid acceptor during its application as a flame retardant to cotton cloth, where it effectively neutralizes acids produced during the process. Esters of ethanolamines are added to wash solutions as antistatic agents for various yarns and fabrics. The esters operate by electrically coupling positive charges on the fibers, neutralizing the charge and its attraction to environmental contaminants.

3. Corrosion Inhibitors and Cement and Concrete Additives.   Ethanol-amines may also be used to form salts that are useful as corrosion inhibitors in metal-working fluids, oil-drilling mixtures, water treatment, or in mixed solvent systems, such as ethylene glycol antifreeze. Equation (7) shows the formation of two inhibitors.

$$HOCH_2CH_2N(H)_2 + H_2SO_4 \rightarrow HOCH_2CH_2N^+(H)_3 + HSO_4^-$$

MEA   Sulfuric Acid

$$(7)$$

$$HOCH_2CH_2N(CH_2CH_2OH)_2 + 3\ H_3PO_4 \rightarrow O = P(OH)_2OCH_2CH_2N(CH_2CH_2OP(O)(OH)_2 + 3\ H_2O$$

TEA   Phosphoric Acid

The inhibitors work in both oil- and water-based formulations by providing a source of phosphate and sulfate functional groups that prevent metal corrosion by penetrating and oxidizing the metal's outside layer. The oxides are resistant to attack from acids generated during everyday use. In oil-based formulations (cutting oils and oil-drilling mixtures), the salts func-tion as emulsifiers, readily accepting corrosive water-soluble materials into the oil. MEA is used when a free base is required and TEA when a salt is required.

The ethanolamines are used as grinding aids, drying accelerators, and as an air entrainer in the manufacture and use of portland cement. During their manufacture, materials in portland cements are burned at elevated temperatures (1450°–2500 °C), and the clinkers are ground to the desired particle sizes after they are processed and blended. TEA is added as a grinding aid to decrease the particle size without increasing the power re-quired to grind the clinkers to the desired size. In general, the particle size determines the speed with which a cement will set and the strength of the concrete. TEA also functions as a drying agent in the preparation of the concrete by helping to absorb moisture from the mixture, causing the ce-ment to dry at a faster rate. TEA is often mixed with either sulfonated aromatic hydrocarbons or with resinous acids as an air entrainer. This property is desirable because cements are often used in areas of frequent freezing and thawing, and it is necessary to add small air pockets to help stop capillary movement of water and prevent undesirable expansion and contraction (Dow Chemical Corp 1988; Howe-Grant 1992; OCC 1995).

4. Pharmaceuticals and Miscellaneous Applications.   Derivatives of ethanolamines are employed in the manufacture of drugs and anesthetics. MEA
is used as a growth accelerator in the production of penicillin, while the
MEA salt of vitamin C is used for intramuscular injection. DEA is used as
a solvent in the manufacture of pharmaceuticals. Aqueous DEA solutions
of drugs such as sulfoxazole are administered intravenously. Methyldietha-
nolamine is used to produce meperidine hydrochloride. The ethanolamines
are also used in the formulation of sulfadiazine for use in the treatment of
burns. The ethanolamines have numerous miscellaneous applications,
which include their use as plasticizers in polyurethanes and intermediates in
the manufacture of glues, adhesives, rubber, and herbicides (OCC 1995)

IV. Acute Toxicity
A. Oral Exposure

The acute oral toxicity of the alkanolamines has been studied in several
laboratory animal species (Table 4). The alkanolamines are only slightly
toxic from single-dose oral exposure. MEA and DEA are of similar degree
in oral toxicity, but both are more toxic than TEA. There does not appear
to be any significant sexual or species differences in acute toxicity with
respect to the alkanolamines.

B. Skin Exposure

The dermal LD_{50} values of MEA, DEA, and TEA from a 24-hr occluded
contact with rabbit skin are given in Table 4. Similar to the pattern of
toxicity seen in animals exposed orally, TEA appears to be the least toxic
from cutaneous exposure among the three alkanolamines.

Table 4. Acute oral and dermal median lethal dose of alkanolamines.

	LD_{50} (g/kg)		
	MEA[a]	DEA[b]	TEA[c]
Oral			
Rat			
Male	1.2–2.5	1.7–2.8	8.4–11.3
Female	1.1–2.7	0.7–1.7	5.5–8.9
Mouse	0.7–15.0	3.3	5.4–7.8
Rabbit	1.0–2.9	2.2	5.2
Guinea pig	0.6	2.0	5.3–8.0
Skin			
Rabbit	1.0–2.5	8.1–12.2	>20

[a]From CIR (1983); BIBRA (1993a); UCC (1988a).
[b]From CIR (1983); BIBRA (1993b); UCC (1988b).
[c]CIR (1983); BIBRA (1990); UCC (1988c).

C. Inhalation Exposure

No inhalation LC_{50} values have been reported for the alkanolamines. However, no mortality was reported for rats exposed for 6 hr to substantially saturated vapor concentrations of MEA, DEA, or TEA generated at room temperature or to a combination of saturated vapor and mist generated at 170 °C. The theoretical saturated vapor concentrations of MEA, DEA, and TEA at room temperature are 520, 0.37, and 0.0047 ppm, respectively. Thus, the LC_{50} of the alkanolamines can be said to be greater than their corresponding saturated vapor concentrations.

D. Primary Irritancy

1. Skin.   Among the three alkanolamines, MEA is the most irritating to rabbit skin. It can cause burns and necrosis to the skin following a 4-hr exposure. MEA is classified as a UN (United Nations) packing group III corrosive substance. Aqueous solutions containing 25%, 50%, or 75% MEA have also been found to be corrosive to rabbit skin (UCC 1982). DEA and TEA, on the other hand, cause only slight and minimal irritation, respectively, to rabbit skin under similar conditions of exposure.

2. Eye.   Variable results have been obtained in eye irritation studies with various formulations and aqueous concentrations of alkanolamines. However, when testing with undiluted material, MEA is generally found to be severely irritating, DEA moderately irritating, and TEA slightly irritating to the rabbit eye.

E. Sensitization

1. Skin.   There have been no animal studies assessing the skin sensitization potential of MEA and DEA, while a number of guinea pig studies with TEA have failed to induce sensitization. Repeated-insult skin patch testing of human volunteers or chemical workers with MEA and DEA have produced negative results, while the results with TEA were equivocal. The overall evidence suggests that both MEA and DEA are not allergenic, and that TEA is at worst weakly allergenic, capable of producing mild skin sensitization in a small fraction of the human population (CIR 1983).

2. Respiratory System.   No studies were found that evaluated the sensitization potential of the alkanolamines to the respiratory tract.

F. Hepatocyte Lesions

According to Blum et al. (1972), an LD_{50} dose of DEA (2.3 g/kg) in the mouse produced abnormal electron microscopic changes in hepatocyte mitochondria and in smooth and rough endoplasmic reticulum (ER). Mitochondrial swelling and loss of matrix were observed 4 hr after administration of DEA. The rough ER was distended, degranulated, and filled with liposomes. Autophagic cytosis of Golgi vesicles of disrupted ER and of microbodies was prevalent. Six hours after administration, the mitochondria in other cells were rounded and densely packed in a fashion similar to that seen in oncocytes. At the end of 24 hr, the architecture of the ER was normal, while the mitochondria were round and rarely annulate. The mitochondrial matrix was densely stained, and granules were present. Widespread changes in mitochondria and ER suggested that cell membranes may have been damaged. The severity and reversibility of the lesion suggested a dose-dependent reaction to DEA. The effect of lower doses of DEA on hepatic phospholipid synthesis, drug metabolism, and mitochondrial structure and function is discussed in the following paragraphs.

V. Pharmacokinetics and Metabolism
A. General

MEA occurs naturally in a group of phospholipids known as phosphatides. This group of complex lipids is composed of glycerol, two fatty acids, and phosphoric acid linked to the hydroxyl group of glycerol and a nitrogenous base such as choline (lecithins, phosphatidylcholine) or MEA (cephalins, phosphatidylethanolamine). The ethanolamine or aminoethanol cephalins are considerably more acidic than are the lecithins because ethanolamine is a weaker base than choline. The amino acid serine is found in place of MEA in a second type of cephalin known as phosphatidylserine. The structures of these materials are given in Fig. 2.

DEA is believed to replace choline, arsenocholine, triethylcholine, and sulfocholine as a component of liver lecithins. According to Artom et al. (1949), these atypical phospholipids are metabolized at a slower rate than their natural analogs, and prolonged administration results in their accumulation in liver. DEA also appears to have a direct or indirect effect on the formation of choline and its utilization in the synthesis of natural cephalins and lecithins. Artom et al. (1949) studied the effect of DEA on the formation of liver phospholipids. According to the studies, a single large dose of either MEA or DEA produced an increase in the formation of liver phospholipids (choline and noncholine). However, when DEA was fed alone in the diet to rats for longer periods of time, a decrease in the formation of choline-containing phospholipids was observed with a marked increase in the noncholine-containing phospholipids. DEA was shown in rats to stimulate lipid phosphorylation (Artom et al. 1949), depress the rate of oxygen absorption by liver tissue, and lower the respiratory quotient (Araksyan 1960), while in mice DEA increases total glycogen, alkaline phosphatase, and weight of the liver, and decreases the total lipid content (Annau and Manginell 1950; Annau et al. 1950). Triethanolamine (TEA) is neither present in nature nor incorporated into natural products when in-

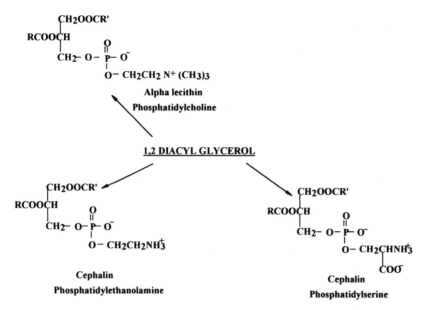

Fig. 2. Structure of lecithins (phosphatidylcholine) and cephalins (phosphatidyl ethanolamine and phosphatidylserine).

gested or absorbed through the lungs or skin. It is eliminated largely in urine as TEA per se (Waechter and Rick 1988).

B. Monoethanolamine

1. Metabolism and Disposition.   Taylor and Richardson (1967) studied the fate of ethanolamine-1,2-C^{14} in the intact rat, tissue slices, and homogenates. Fifty-four percent of the dose was found in liver, spleen, kidneys, heart, brain, and diaphragm, and 11.5% was accounted for as $^{14}CO_2$ 8 hr after intraperitoneal administration. The radioactivity in tissues was found distributed in lipid, amino acid, organic acid, and sugar fractions. Approximately 85% of the tissue radioactivity were found in the lipid fraction. The liver was the most active tissue, followed by the heart and brain. Figure 3 gives the metabolic pathway for the incorporation of MEA into liver phosphatidylethanolamines via phosphorylethanolamine and CDP-ethanolamine (cytidine-5′-diphosphoethanolamine) (Sundler 1973).

A pathway leading to the formation of respiratory CO_2 from ethanolamine was suggested by Taylor and Richardson (1967) to incorporate the formation of glycoaldehyde via transamination or deamination reactions involving a pyridoxine-dependent pathway. Glycoaldehyde, however, was not identified as an intermediate in the oxidation of ethanolamine by rat liver slices containing mitochondrial amine oxidase. Sprinson and Weliky (1969) postulated a metabolic pathway leading to $^{14}CO_2$ in the rat using

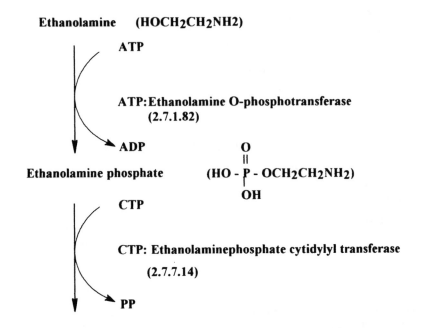

Ethanolamine   **(HOCH$_2$CH$_2$NH$_2$)**

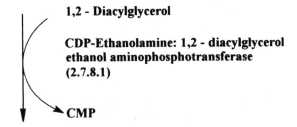

Phosphatidylethanolamine (Cephalin)

Fig. 3. Metabolism of ethanolamine to phosphatidylethanolamine. *ATP*, Adenosine-5′-triphosphate; *ADP*, adenosine-5′-diphosphate; *CTP*, cytidine-5′-triphosphate; *CDP*, cytidine-5′diphosphate; *PP*, pyrophosphate; *CMP*, cytidine-5′-monophosphate. Enzyme code number in (parentheses); systematic name given above code numbers. (Numbers and names from *Enzyme Nomenclature 1992*, Academic Press, San Diego, CA.)

[1-D$_2$, 2-^{14}C]ethanolamine via ethanolamine-*O*-phosphate. Activation of hydrogen on the amino carbon atom by pyridoxal phosphate followed by the elimination of inorganic phosphate gives an eneamine that is hydrolyzed to acetaldehyde, ammonia, and pyridoxal phosphate. The conversion of phosphorylated ethanolamine to acetaldehyde restricts the amount of ethanolamine being incorporated into phospholipids via CDP-ethanolamine.

Babior (1969) studied the conversion of ethanolamine to acetaldehyde and ammonia by ethanolamine deaminase, a B_{12}-coenzyme-dependent enzyme. The reaction involved (1) the intermolecular migration of one hydrogen atom from the carbinol carbon atom to the adjacent amino carbon atom and (2) the transfer of the amino group to the carbinol carbon. Ammonia is lost from the newly formed 1-amino alcohol to form acetaldehyde, which is further metabolized to carbon dioxide.

Enzymes Involved in Phosphorylation and Transfer Reactions. Phosphorylation of ethanolamine [adenosine triphosphate (ATP): ethanolamine phosphokinase] (EC 2.7.1.82) occurs exclusively in the supernatant, while phosphorylation of ethanolamine from sphingolipid (composed of two nitrogenous bases, choline and sphingosine, fatty acids, and phosphoric acid) is catalyzed by a membrane-bound enzyme. The CDP:ethanolamine-phosphate cytidyltransferase (EC 2.7.7.14) is located in the particle-free supernatant, whereas ethanolaminephosphotransferase (EC 2.7.8.1) is a microsomal enzyme (see Fig. 3).

Metabolic Rate Constants: Liver and Brain. Table 5 gives the size of the metabolic pool in rat liver for the ethanolamine-containing compounds; Fig. 4 gives the rate of conversion of intraportally injected $[2\text{-}^3\text{H}]$ ethanolamine into phosphorylethanolamine, CDP-ethanolamine, and phosphatidylethanolamines (Sundler 1973). The specific activity of CDP-ethanolamine was about twice that of phosphorylethanolamine, suggesting that phosphorylethanolamine formed from exogenous ethanolamine is not completely mixed with other liver pools of phosphorylethanolamine. The distribution of radioactive ethanolamine among the phosphatidylethanolamines indicated that it had to be incorporated mainly via CDP-ethanolamine and not by base exchange. The rate of synthesis of phosphatidylethanolamines from CDP-ethanolamine was of the order of 0.06–0.08 μmole/min per liver (Sundler 1973).

Previous investigations have shown that when $[2\text{-}^{14}\text{C}]$ethanolamine was injected intracerebrally, it was rapidly phosphorylated at a rate of 0.14

Table 5. Pool size in rat liver for ethanolamine-containing compounds.

Compound	Pool size (μmol/liver)
Ethanolamine	1.09 ± 0.14
Phosphorylethanolamine	3.8 ± 1.08
CDP-Ethanolamine	0.239 ± 0.053
Phosphatidylethanolamines	80.59 ± 8.82

From Sundler (1973), with kind permission from Elsevier Science-NL, Amsterdam, The Netherlands.

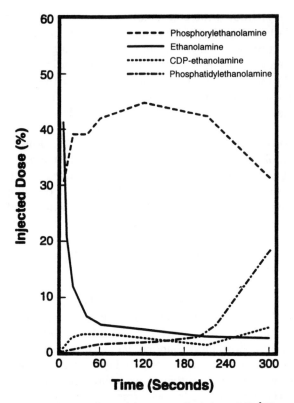

Fig. 4. Conversion of intraportally injected [2-³H]-
ethanolamine into phosphorylethanolamine, CDP-
ethanolamine, and phosphatidylethanolamine. (Re-
drawn from Sundler 1973.)

μmole/g of brain per hour (Ansell and Spanner 1966). There was also a
large incorporation into ethanolamine phospholipids. The rate of turnover
of the phosphatidylethanolamines in liver was not studied by Sundler
(1973). Studies by Horrocks (1969), however, involving the ethanolamine
phosphoglycerides of mouse brain myelin and microsomes indicate that
some or all brain ethanolamine phosphoglycerides (diacyl and alkyl acyl
glycerylphosphorylethanolamines) turn over rapidly, with an apparent half-
life of less than 3 d.

The only known pathway for the de novo synthesis of phosphatidylcholine (PC) is by methylation of phosphatidylethanolamine (PE) to form PC.
Wise and Elwyn (1965) estimated that this pathway may provide choline in
amounts equivalent to the dietary intake of about 13 μmol d^{-1} g^{-1} of liver.
Table 6 shows the amount of administered radioactivity found in PE and
PC during a 5-hr period (Tinoco et al. 1970). Maximum incorporation
in the PE fraction was attained within 10 min after administration. This
percentage decreased slowly with the progressive formation of labeled PC.

Table 6. Incorporation of [1,2-^{14}C]ethanolamine into phosphatidylethanol-amine (PE) and phosphatidylcholine (PC) in female rat liver.

Time (min)[a]	Phosphatidylethanolamines (% of dose)	Phosphatidylcholines (% of dose)
10	27.1 ± 1.7	0.7 ± 0
20	26.0 ± 1.7	1.9 ± 0.7
60	26.7 ± 1.0	4.7 ± 0.1
300	22.1 ± 0.6	9.8 ± 1.3

[a]Three rats were analyzed separately for each time.

From Tinoco et al. (1970), with permission from American Oil Chemists' Society.

Fatty Acid Compositon of Phospholipids.   Table 7 gives the distribution of radioactive ethanolamine among phosphatidylethanolamine fractions of different unsaturated fatty acids. According to the data, the specificity of CDP-ethanolamine:1,2-diacyl-glycerol ethanolamine phosphotransferase (EC 2.7.8.1) was less sensitive to the chain length of the fatty acid in the

Table 7. Distribution of [^{14}C-2]ethanolamine among phosphatidylethanolamine fractions of different unsaturated fatty acids after intraportal injection.

Phosphatidyletha-nolamine fraction	Time after injection (sec)					
	10	20	40	60	210	300
Monoenoic[a]	8.70	9.69	9.27	9.24	10.80	9.39
Dienoic[b]	13.26	15.44	13.25	12.47	14.27	14.27
Trienoic[c]	4.19	7.83	6.34	6.54	4.14	6.73
Tetraenoic[d]	14.37	15.23	16.47	14.65	19.17	15.56
Pentaenoic		5.53	5.46	4.61		5.12
Pentaenoic + hexaenoic	59.49					
Hexaenoic		46.28	49.20	52.50		48.94

[a]e.g., Oleic acid; octadec-9-enoic acid; $CH_3(CH_2)_5CH=CH(CH_2)_7COOH$.

[b]e.g., Linoleic acid; octadeca-9,12-dienoic acid; $CH_3(CH_2)_3(CH_2CH=CH)_2(CH_2)_7COOH$.

[c]e.g., Linolenic acid; octadeca-9,12,15-trienoic acid; $CH_3(CH_2CH=CH)_3(CH_2)_7COOH$.

[d]e.g., Arachidonic acid; eicosa-5,8,11,14-tetraenoic acid; $CH_3(CH_2)_3(CH_2CH=CH)_4(CH_2)_3$ COOH.

The data are given as percentage of total phosphatidylethanolamine radioactivity and represent mean values from 2-4 rats except for the values at 10 and 210 sec from single rats.

Data from Sundler (1973), with kind permission from Elsevier Science-NL, Amsterdam, The Netherlands.

diacylglycerol than it was to their degree of unsaturation. The chemical structures and common names of several unsaturated mono-, di-, tri-, and tetraenoic acids found in nature are listed in Table 7, with linolenic and arachidonic acids being the most common trienoic and tetraenoic acids in phospholipids.

2. Pharmacokinetics and Percutaneous Absorption. Sun et al. (1996) conducted an *in vitro* dermal penetration study using ^{14}C-MEA (15 mCi/mmol), freshly excised human, rat, mouse, and rabbit skin, and the skin permeation system described by Holland et al. (1984). Skin disks were prepared according to the method described by Kao et al. (1983). In this procedure, animal hair is clipped prior to harvesting the skin. Fat and connective tissues are gently removed and the skin placed in a petri dish containing a minimum essential medium (MEM) described by Eagle (1959); 4 mg/cm^2 of ^{14}C-MEA (5–10 μCi/skin) was then applied either undiluted or diluted in water to 1.77 cm^2 of skin. A layer of gauze, cut to the size of the skin, was placed over the topically applied dose to simulate the application procedure used in a developmental toxicity study (Liberacki et al. in press) and to ensure skin contact. The amount of ^{14}C penetrating the skin was determined by dividing the disintegrations per mmule (dpm) in the effluent by that in the topically applied dose. The steady-state penetration rate was used to calculate the permeability constant by dividing the slope of the line (mg cm^{-2} hr^{-1}) by the initial concentration (mg/cm^3) of MEA applied to each skin disk (Bronaugh et al. 1982, 1986). The results of the study using aqueous MEA are given in Table 8. Permeation (k_p) was less using undiluted MEA (i.e., 1.21 for undiluted vs. 7.55 for diluted studies involving the mouse) in the tested skin. Long lag times or delays in absorp-

Table 8. *In vitro* percutaneous absorption of ^{14}C-MEA (aqueous dose): comparison across species.

Species	k_p (cm/hr $\times$ 10^{-4})	Absorption rate (mg cm^{-2}hr^{-1})	Percentage of dose absorbed (6 hr)
Rat	0.53 (2.93 $\times$ 10^{-3})	0.0117	1.32
Mouse	7.55 (4.23 $\times$ 10^{-2})	0.1694	24.79
Rabbit	1.13 (6.3 $\times$ 10^{-3})	0.0253	1.84
Human	0.43 (2.4 $\times$ 10^{-3})	0.0097	1.11

k_P, Permeation.
k_p, mg cm^{-2} hr^{-1} $\div$ mg/cm^3; applied dose = 4 mg/cm^3 (32 μL applied, 22% w/w), 1.77 cm^2 area.
k_p values in parentheses calculated by authors in this review.
Total recovery: rat, 72.5%; mouse, 90.8%; rabbit, 82.4%; human, 69.7%.
Information from Sun et al. (1996), courtesy of Marcel Dekker, Inc.

tion (1.0–2.0 hr) (Fig. 5) were encountered with human skin, with little or no delay encountered with mouse or rat skin.

C. Diethanolamine

1. Metabolism and Disposition

Single Bolus Studies: Intravenous and Oral.   Forty-eight hours after the intravenous administration of ^{14}C-diethanolamine (7.5 mg/kg) to the rat, 28% and 1% of the dose was excreted in urine and feces, respectively. The remaining portion of the administered dose was retained in tissues, with the highest concentrations located in the liver and kidney. The tissue to blood ratios for the liver and kidney were 150, while the ratios for lung and spleen were 35–40, for heart approximately 20, and in other tissues less than 10 (Mathews and Jeffcoat 1991). Analysis of liver and urine indicated that diethanolamine was not metabolized to CO_2 and other materials.

Mathews et al. (1995) studied the metabolism and incorporation of DEA in phospholipids. Adult male Fischer-344 rats were administered 7 mg of ^{14}C-DEA/kg intravenously and orally in water. Less than 30% of the orally administered dose was eliminated over a 48-hr period in urine and feces;

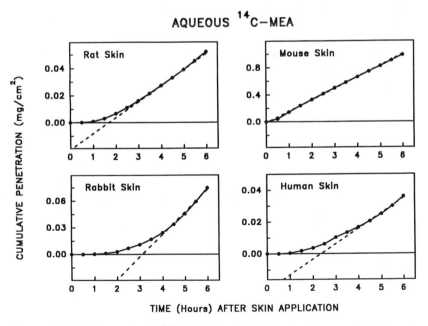

Fig. 5. Rate of absorption of ^{14}C-MEA through isolated mouse, rat, rabbit, and human skin. (From Sun et al. 1996, Journal of Toxicology — Cutaneous and Ocular Toxicology, courtesy of Marcel Dekker, Inc.)

27% of the orally administered dose was found in liver after 48 hr, with lesser amounts being present in other tissues. These results were similar to those obtained in intravenous rat studies conducted by Waechter et al. (1995) using uniformly labeled ^{14}C-diethanolamine in which urinary elimination of DEA per se accounted for 25% and 36% of 10 and 100 mg/kg doses, respectively. An average of 64.1% and 51.5% of the administered radioactivity was recovered 96 hr post-administration in the tissues of animals dosed at levels of 10 and 100 mg/kg, respectively. The majority of the radioactivity retained was found in the carcass (34.6% and 28.2%, 10 and 100 mg/kg), liver (20.9% and 17.1%, 10 and 100 mg/kg), and kidneys (7.2% and 4.9%, 10 and 100 mg/kg). Less than 1% of the radioactivity was retained by the brain, fat, heart, and stomach, while approximately 5% was retained by skin. Kidneys contained the highest concentration (26 and 199 μg/g, 10 and 100 mg/kg), with the liver containing slightly less diethanolamine (15 and 137 μg/g, 10 and 100 mg/kg).

Mathews et al. (1995) extracted liver, brain, and blood phosphate-buffered saline (PBS) homogenates individually with $CHCl_3$:MeOH to remove and characterize the residues in these tissues. The majority of the radioactivity (87%–89%) remained in the water phase, with 6%–9% partitioning into the chloroform phase. High performance liquid chromatography (HPLC) analysis of the water phase revealed the presence of DEA (~80%), N-methyl-DEA, N,N-dimethyl-DEA, and their combined phosphates (15%). Incubation of the phosphates with alkaline phosphatase released DEA, N-methyl DEA, and N,N-dimethyl DEA (95%, 2.0%, and 2.0%, respectively).

DEA was also the major component found in the aqueous phase of brain homogenates. No methylated DEA metabolites were found, and phosphorylated DEA was the only metabolite found. The chloroform extracts of the liver PBS homogenate contained phosphatidylethanolamine (PE) and phosphatidylcholine (PC) by HPLC analysis. Incubations of HPLC fractions with phosphatase yielded DEA in the PE fraction and N-methyl DEA (15%) and N,N-dimethyl DEA (85%) in the PC fraction. Analysis of the chloroform extracts from brain under similar conditions yielded primarily DEA. In a separate study by Mathews et al. (1995), the combined phospholipids (sphingomyelin and phosphatides) were subjected to the action of spingomyelinase followed by phosphatase. The study showed that 30% of the DEA-derived phospholipids were ceramides (sphingolipids) and 70% were phosphoglycerides.

Repeated-Dose Studies: Oral.   Mathews et al. (1995) determined the distribution of ^{14}C-DEA in selected tissues taken 72 hr after the final dose of an 8-wk oral study. Analysis of liver showed that 97% of the radioactivity was present in the aqueous phase and 2% in the chloroform phase, while in brain the aqueous phase contained 77% of the radioactivity and the chloroform phase 21% (compared to 6% in the single-dose study, 48 hr post-administration). DEA was the major component in the aqueous phase of both tissues. The organic phase from liver chromatographed as radiolabeled phosphatidylcholine (PC) and phosphatidylethanolamine (PE). Enzymatic hydrolysis of the PC and PE fractions liberated *N*-methyl DEA (15%) and *N,N*-dimethyl DEA (85%) in the PC fraction and DEA (~100%) in the PE fraction. Under similar analytical conditions, the chloroform phase from brain contained >97% DEA.

Twenty percent of the radioactivity in blood was present in phospholipids and the remainder as water-soluble material in the single-dose and 8-wk repeated-dose studies at 48 and 72 hr post-administration. DEA in the lipid fractions (80%) was methylated in the single-dose studies, and practically all DEA was methylated in the repeated-dose study. The reason for the retention of DEA per se in tissues is related to its metabolic stability (i.e., metabolism to CO_2 and natural products) and the rate at which DEA is phosphorylated and incorporated into lipids. Mathews et al. (1995) incubated human liver slices with radiolabeled DEA (1.0 mM) for periods of 4 and 12 hr. Small amounts of the DEA taken up by slices (11%–29%) were incorporated into phospholipids (7.2%–14.2%) during incubation. The majority of the radioactivity was identified as DEA.

2. Pharmacokinetics

Intravenous: Single Bolus.   In a recent metabolism/pharmacokinetic study conducted by Waechter et al. (1995), uniformly labeled ^{14}C-diethanolamine was administered intravenously (10 and 100 mg/kg) in physiological saline to Sprague–Dawley rats. The intravenous dose of 100 mg/kg was tolerated by the rats without any signs of toxicity. Ten mg/kg was selected as the lowest dose on the basis of lowest-observed-effect level (LOEL) in a 90-d oral study (Hejtmancik et al. 1988). Animals were housed in all-glass metabolism cages to facilitate the collection of urine and feces. Blood samples were collected at 5, 10, 15, and 30 min and at 1, 2, 4, 6, 12, 24, 36, 48, 60, and 72 hr post-administration.

Animals were sacrificed after 96 hr and the tissues analyzed for ^{14}C activity. Peak blood (red cells and plasma) concentrations occurred 5 min after the administration of the dose. Elimination from blood (plasma and red cells) was biphasic (α and β phases), with plasma half-lives of 9.2 and 258 hr, respectively, for the 10 mg/kg dose. Similar half-life values were found for the red cells. Longer α phase half-lives, 16.3 hr, and shorter β phase half-lives, 206 hr, were reported for the 100 mg/kg doses. Urinary elimination accounted for 25% and 36% of the 10 and 100 mg/kg doses, respectively. The urinary concentrations of DEA were determined to be 18.9, 6.35, 1.92, and 2.14 μg/g for the 0–12, 36–48, 60–72, and 84–96 hr collections, respectively, for a 10 mg/kg dose level. A disproportionate increase (40-fold higher) in DEA urinary concentrations (708 μg/g) was observed with 0–12 hr urine samples from animals administered 100 mg/kg.

Percutaneous Absorption.   An *in vitro* absorption/permeation study was conducted by Sun et al. (1996) using ^{14}C-DEA and human, rat, mouse, and rabbit skin. The procedure used was identical to the one previously described for MEA. The permeability (k_p, cm/hr) of aqueous DEA was similar in rat and human skin but greater in mouse and rabbit skin (see Table 9).

In single-dose percutaneous absorption studies using the rat, ^{14}C-DEA was topically applied to 2-cm^2 areas at doses of 2.1, 7.6, or 27.5 mg/kg in 95% ethanol and left on the skin for 48 hr (Mathews and Jeffcoat 1991). Percutaneous absorption varied with dose, ranging from 3% at a dose of 2.1 mg/kg to 16% at a dose of 27.5 mg/kg. Permeation constants shown in Table 10 ranged from 6.0×10^{-4} to 3.36×10^{-3} cm/hr.

In a more recent rat single-dose percutaneous absorption study conducted by Waechter et al. (1995), ^{14}C-labeled DEA was applied topically to 19.5 cm^2 of clipped back skin (~ 20 mg/cm^2, 1500 mg/kg). The treated site was occluded with nonabsorbent gauze and Saran® film. Animals were individually held for 48 hr in all-glass metabolism cages to facilitate the complete collection of urine and feces. Wrappings were removed after 6 hr, and the treated skin area on 50% of the animals and all wrappings were washed to remove residual DEA. Skin washings and wrappings were analyzed for ^{14}C-DEA. Blood samples were collected via indwelling jugular catheter at 5, 10, 15, 30 min and 1, 2, 4, 6, 12, 24, and 36 hr post-administration. In the case of unwashed animals, the majority of the dose was recovered in wrappings (80%) and skin at the dose site (3.6%). Six-hour skin washings from washed animals contained 26% and washings from wrappings 57.8% of the applied dose. Absorbed ^{14}C-DEA was determined by analysis of urine, feces, and tissues (excluding skin and skin washings). Unwashed animals absorbed 1.4% and washed animals 0.64% of the dose. The majority of the radioactivity was associated with the car-

Table 9. *In vitro* percutaneous absorption of ^{14}C-DEA (aqueous dose): comparison across species.

Species	k_p (cm/hr $\times 10^{-4}$)	Absorption rate (mg cm^{-2} hr^{-1})	Percentage of dose absorbed (6 hr)
Rat	0.60 (1.15×10^{-3})	0.0230	0.56
Mouse	7.62 (1.47×10^{-2})	0.2944	6.68
Rabbit	3.42 ($6.6 \ \times 10^{-3}$)	0.1322	2.81
Human	0.34 (0.64×10^{-3})	0.0127	0.23

k_p, mg cm^{-2} hr^{-1} $\div$ mg/cm^3; dose $= 20$ mg/cm^3 (35 μL of 37% w/w solution), applied to 1.77 cm^2.

k_p values in parentheses calculated by authors in this review using 20 mg/cm^3.

Total recovery: rabbit, 84.3%; rat, 83%; mouse, 80.7%; human, 77.3%.

Information from Sun et al. (1996), courtesy of Marcel Dekker, Inc.

Table 10. *In vivo* percutaneous absorption of ^{14}C-DEA in rats and mice.

	Single dose, 48-hr duration					Multiple dose	
						3 Day	6 Day
Male F-344 rats							
Applied dose							
mg/kg	2.1[a,c]	7.6[a,c]	27.5[a,c]	1500[b,d]	1500[b,g,e]	1500[b,f,g]	1500[b,f,g]
$\mu g/cm^3$	187.5	679.4	1725.6	19720	19990	16128	15828
Absorption							
$\mu g\ cm^{-2}\ hr^{-1}$	0.113[h]	1.48[h]	5.8[h]	45.0[h]	21.0[h]	70.4[i]	136.9[i]
k_p, cm/hr[j]	6.0×10^{-4}	2.2×10^{-2}	3.36×10^{-3}	2.28×10^{-3}	1.05×10^{-3}	4.36×10^{-3}	8.65×10^{-3}
Mice B6CF1							
Applied dose							
mg/kg			81.1[a]				
$\mu g/cm^3$			1622				
Absorption							
$\mu g\ cm^{-2}\ hr^{-1}$			19.9[j]				
k_p, cm/hr[j]			2.1×10^{-2}				

[a] Absorption rates calculated from percutaneous absorption sutides of Mathews and Jeffcoat (1991).
[b] Absorption rates calculated from percutaneous absorption studies of Waechter et al. (1994).
[c] Dose applied to 2.0 cm², left on skin for 48 hr.
[d] Dose applied to 19.5 cm² of skin, wrappings removed after 6-hr exposure, skin unwashed.
[e] Dose applied to 19.5 cm² of skin, wrappings removed after 6-hr exposure, skin washed.
[f] Repeat dose applied (for 3 and 6 d) to 19.5 cm² of skin followed by ^{14}C-labeled DEA (1500 mg/kg).
[g] Radiolabeled dose left on skin for 48 hr.
[h] Absorption rate ($\mu g\ cm^{-2}\ hr^{-1}$) based on 6-hr data.
[i] Absorption rate ($\mu g\ cm^{-2}\ hr^{-1}$) based on 48-hr data.
[j] Calculated by dividing the rate ($\mu g\ cm^{-2}\ hr^{-1}$) by $\mu g\ cm^3$ applied.
[k] Weight of mice estimated to be 20 g; dose applied to 1.0 cm², left on skin for 48 hr.
Values in table calculated by authors.

cass, liver, or kidneys. Little if any radioactivity was recovered in urine (0.11%) and feces. Detectable but nonquantifiable radioactivity was present in plasma and red cells. The absorption and permeation rates shown in Table 10 were calculated on the basis of the applied dose and all absorption occurring during a 6-hr period.

The absorption rates (μg cm^{-2} hr^{-1}) determined from unwashed animals in the Mathews and Jeffcoat (1991) study were used as the first three points in the linear plot shown in Fig. 6, with the fourth data point being taken from the Waechter et al. (1995) study. The resulting plot clearly shows that a 100-fold increase in the concentration of DEA applied to skin (187.5–19,720 μg/cm^3) resulted in a 450-fold increase in the rate of absorption (0.113–45.0 μg cm^{-2} hr^{-1}) of DEA. Several confounding factors, such as the use of gauze bandages, differences in the age of the animals, and the use of ethanol, were present in these studies. These factors, however, do not appear to have adversely affected the linearity of the data.

In a repeated percutaneous study performed by Waechter et al. (1995), 1500 mg/kg of nonradiolabeled DEA was topically administered to the clipped back of rats once per day, 6 hr/d for 3 or 6 d. The nonradiolabeled dosing solution was applied to the dorsal skin under a 2 × 2 in. gauze square (25.0 cm^2) and occluded with Saran® film to reduce oral contact with the material. ^{14}C-DEA (1500 mg/kg) was then applied to the skin of the animals dosed for 3 or 6 consecutive days. The radiolabeled dose was left in contact with the skin for 48 hr prior to removing the wrappings and washing the treated skin and wrappings.

Animals in the 3-d and 6-d groups absorbed 21% and 41% of the applied

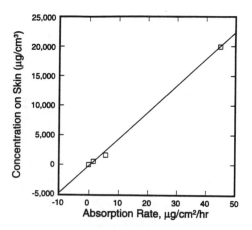

Fig. 6. Percutaneous absorption of ^{14}C-DEA (μg cm^{-2} hr^{-1}) through the back skin of the rat as a function of single topical dose (μg/cm^3). (Data taken from Table 10.)

dose, respectively. The absorbed dose included the radioactivity recovered in the carcass, excreta, and internal organs. Dosed skin and skin remote from the dose site were excluded. A majority of the topically applied doses (70% and 47% for the 3-d and 6-d dose groups, respectively) were recovered in the wrappings used to occlude the dose site. Carcass, liver, or kidney contained the majority of radioactivity. Less than 0.3% of the administered dose was found in brain, fat, or heart in either dose group. Urine from the 3-d and 6-d groups contained 4.3% and 13% of the applied dose, respectively. The overall recovery was good (96%–97%) for the two groups.

Preexposure to DEA resulted in a 1.5- to 3.0-fold increase in the absorption rate (μg cm^{-2} hr^{-1}). This increase is small compared to the 450-fold increase produced by increasing the concentration applied to skin. The absorption rates and permeability constants calculated from the percutaneous absorption data of Waechter et al. (1995) are given in Table 10. Figure 7 shows the effects of single topical applications and repeated applications on absorption.

Four $B_6C_3F_1$ mice were each topically administered ^{14}C-DEA (81.1 mg/kg) by Mathews and Jeffcoat (1991). The dose was applied to 1.0 cm^2 of clipped back skin. A metal appliance was glued onto the animal's back prior to returning the animal to its cage. Animals were housed separately in all-glass metabolism cages to facilitate the complete collection of urine, feces, and volatiles in expired air over a 48-hr period. Blood was collected

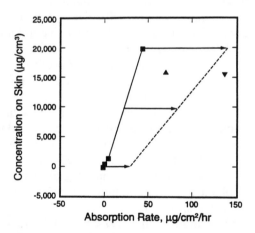

Fig. 7. Percutaneous absorption of ^{14}C-DEA (μg cm^{-2} hr^{-1}) through the back skin of the rat after a single topical application (shown in Fig. 6, *squares*), and following repeated topical applications of DEA at 3 d (*upward-pointing triangle*) and 6 d (*downward-pointing triangle*), as a function of topical dose (μg/cm^3). (Data taken from Table 10.)

from each animal at sacrifice along with treated skin, appliance washings, adipose tissue, skin, kidney, liver, spleen, heart, lungs, and brain. Tissues, excreta, and washings were analyzed for radioactivity. DEA was absorbed (59% of applied dose) through back skin at the rate of 19.9 μg cm^{-2} hr^{-1}. A k_p value of 0.021 cm/hr was calculated based on this rate (see Table 10). Radioactivity present in tissue after 48 hr was found largely in kidney (6.04%), liver (20.36%), and muscle (15.79%).

3. Inhibition and Effects

Phospholipid Synthesis.   Of the three alkanolamines studied (MEA, DEA, and TEA), only MEA is found in nature as a constituent in phospholipids. TEA is not found in nature and when administered is readily eliminated from the body and not incorporated into phospholipids. DEA, on the other hand, is closely related to MEA and interacts in phospholipid metabolism by competing with ethanolamine and choline (Barbee and Hartung 1979a; Chojnacki and Korzybski 1963; Morin 1969; Welch and Landau 1942; Wells and Remy 1961) in the *in vitro* synthesis of phosphatidylcholine and phosphatidylethanolamine in liver tissue. According to a number of investigators (Jungalwala and Dawson 1970a,b; McMurray and Dawson 1969; Schneider 1963; Wilgram and Kennedy 1963), the primary site of phospholipid synthesis is located on the endoplasmic reticulum.

Barbee and Hartung (1979a) studied the *in vitro* inhibition of phospholipid synthesis by DEA. In practice, they incubated 100, 300, 1000, and 3000 μM DEA, individually, with a 1000-g liver fraction for 1 hr at 37 °C and followed the rate of incorporation of radioactive ethanolamine or choline into phosphatidylethanolamine or phosphatidylcholine. According to a Lineweaver–Burke plot for the incorporation of choline into phosphatidylcholine, K_m decreased with increasing DEA concentrations, while V_{max} remained constant. The K_i for DEA inhibition of phosphatidylcholine synthesis was 2600–2900 μM. In the incorporation of ethanolamine into phosphatidylethanolamine, K_m decreased with increasing DEA concentrations and V_{max} increased. The K_i value was estimated to be 2600–2900 μM. The K_m, V_{max} value for the incorporation of DEA into phospholipids was determined to be 11,600 μM and 21.0 nmol/mg of protein per hour. The K_m values for phosphatidylcholine and phosphatidylethanolamine synthesis were 75.5 and 53.5 μM, respectively.

According to Barbee and Hartung (1979a), a single oral dose of 250 mg/ kg DEA did not inhibit the *in vitro* synthesis of choline and ethanolamine phospholipids, while daily administration of DEA in drinking water (320 mg kg^{-1} d^{-1}) over a 3-wk period inhibited the incorporation of choline and ethanolamine into phospholipids. Daily administration of DEA appears to be required to raise the *in vivo* liver concentration 150-fold (11,600 μM) above that of ethanolamine (53.5 μM) or choline (75.5 μM). Barbee and Hartung (1979a) determined the half-lives for the elimination of ^{14}C-DEA phospholipids from liver and kidney to be 3.5 and 4.2 d, respectively. The half-lives for [^{3}H]choline-labeled phospholipids were 1.7 d in liver and 2.1 d in kidney.

Drug-Metabolizing Enzymes.   Foster (1971) found that repeated dosing with DEA inhibited the activity of hepatic microsomal drug-metabolizing enzymes similar to the *in vivo* inhibition of the formation of phosphatidylethanolamine and phosphatidylcholine by DEA. The reason for this is unclear but may be related to the formation of foreign phospholipids incapable of carrying out the function of natural cellular constituents or competition between drugs and DEA in the biomembrane.

Mitochondrial Function and Structure.   Barbee and Hartung (1979b) studied the effects of DEA on hepatic mitochondrial function and structure using isolated hepatic mitochondria from control rats and from rats administered DEA acutely and repeatedly. The *in vitro* and acute *in vivo* effect of DEA on hepatic mitochondria was assayed at 5 mM (5000 μM) and 490 mg/kg, respectively. The repeated-dose *in vivo* effect was measured using mitochondria from rats administered neutralized DEA in drinking water for 2 and 5 wk (42 mg kg^{-1} d^{-1}), for 1, 3, and 5 wk (160 mg kg^{-1} d^{-1}), and for 1 d, 3 d, and 1, 2, 3, and 5 wk (490 mg kg^{-1} d^{-1}).

The effect of DEA on respiratory rate, acceptor control ratio (State 3 activity/State 4 activity), and adenosine diphosphate : oxygen ratio (ADP/O) for hepatic mitochondria was determined using an oxygen electrode (Estabrook 1967). The assay was initiated by the addition of 50 μL mitochondrial suspension and 200 nmol ADP in 0.15 M sucrose, 20 mM potassium chloride, 20 mM phosphate (monobasic), 5 mM magnesium chloride, and 100 mM succinate (Brabec et al. 1974). The addition of ADP causes an immediate increase in the rate of oxygen utilization (State 3, nmol O_2 mg^{-1} min^{-1} of protein) and the rate of phosphorylation of ADP. State 4 (nmol O_2 mg protein min^{-1}) occurs at the time, ADP becomes limiting, and the rate of oxygen consumption decreases. State 4 activity is defined as respiration in the absence of ADP. The ADP/O ratio was calculated by dividing the nannomoles of ADP phosphorylated by nannoatoms of oxygen consumed during each interval at State 3. DEA did alter the mitochondrial variables (State 3 activity, State 4 activity, acceptor control ratio, and ADP/O ratio) in the *in vitro* (5 mM) and *in vivo* (acute study, 490 mg/kg) studies. Repeated-dose administration, however, resulted in (1) an increase in State 4 activity at all dose levels following 2 wk of administration and (2) a decrease in the acceptor control ratio at dose levels of 160 and 490 mg kg^{-1} d^{-1} after 1 wk of administration. No significant differences in the ADP/O ratio occurred at any time or dose level.

In addition to the foregoing studies, the permeability of the inner and outer mitochondrial membranes was measured to explain the increase in State 4 activity. The method and duration of administration and dose of

DEA (490 mg kg^{-1} d^{-1} in drinking water for 2 wk) were kept constant for both investigations. Outer mitochondrial membrane permeability was measured according to a modification of the method of Wattiaux-DeConnick and Wattiaux (1971) involving the use of cytochrome c. According to Barbee and Hartung (1979b), cytochrome c readily migrates through the membrane when it is sufficiently altered by detergents or toxic substances. The permeability of the inner membrane was determined following the method of Byard et al. (1975). Mitochondria were isolated and assayed for oxygen consumption using nicotinamide adenine dinucleotide (NADH) in place of succinate as substrate. The results were expressed as the difference in respiration between an excess of NADH and ADP compared to no added NADH and an excess of ADP, with units given as nmol O_2 min^{-1} mg^{-1} of protein. No increase in permeability toward NADH and cytochrome c over controls was observed in these studies. The possibility exists that mitochondria are more permeable to smaller ions. The mitochondria from DEA-treated animals are spherical and appear larger than those from control animals under the electron microscope. Barbee and Hartung (1979b) were unable to define the mechanism by which DEA alters hepatic mitochondrial function and structure, but they presumed that the effects were related to the involvement of DEA with phospholipid metabolism.

D. Triethanolamine

1. Metabolism.   The metabolism of ^{14}C-TEA in the mouse was studied by Waechter and Rick (1988) as part of their intravenous and dermal pharmacokinetic studies. Urine from treated animals was chromatographed on an ion-exchange column. The radioactivity eluted as one peak and cochromatographed with authentic standards of TEA. Mass spectral analysis of freeze-dried fractions of the eluant from the ion-exchange column produced a spectrum virtually identical to that of the TEA standard, while analysis by gas chromatography/mass selective detector (GC/MSD) of selected fractions eluting off the ion-exchange column produced no peaks eluting at the retention times of MEA or DEA.

2. Pharmacokinetics

Intravenous: Single Bolus to Mice.   The fate of a 1.0 mg/kg iv dose of ^{14}C-TEA was investigated by Waechter and Rick (1988) in the mouse. Urine, feces, and expired CO_2 were collected using Roth cages at 12 and 24 hr post-administration. Blood samples were collected from all animals via orbital sinus puncture in groups of three at 0.083, 0.167, 0.5, 1, 2, 4, 6, 12, and 24 hr after administration. Radioactivity in excreta, CO_2-trapping solution, cage washings, liver, kidneys, skin, and carcass was determined. Urine was analyzed for parent and metabolites using ion-exchange chromatography and mass spectrometry. The radioactivity was eliminated from blood in a first-order biphasic manner. The half-life of the rapid (or α) phase was 0.58 hr, while the half-life of the slow (or β) phase was 10.2 hr using the method of residuals (Gibaldi and Perrier 1982).

Percutaneous: Mice and Rats.   Topical applications by Waechter and Rick (1988) of 1000 mg/kg ^{14}C-TEA in acetone to 1.0 cm^2 of mouse back skin resulted in ^{14}C-blood levels roughly 1000-fold higher than those found for mice intravenously administered 1.0 mg/kg. The half-life for the elimination of TEA was estimated to be 9.7 hr from the simultaneous absorption–elimination curve. Topical applications of 2000 mg/kg ^{14}C-TEA to 2.0 cm^2 of skin (no ring) resulted in peak radioactivity in blood 2 hr post-administration, with a half-life of absorption and elimination of 10.2 min (0.17 hr) and 18.6 hr, respectively. The concentrations of radioactivity in blood were fivefold greater for the 2000 mg/kg dose than were found for the 1000 mg/kg dose. Percutaneous absorption was more variable in mice equipped with glass rings (1.75 cm^2), with an estimated absorption of 4.2 hr versus 10.2 min without the glass ring. The difference appears to result from oral ingestion by the mouse. Terminal blood concentrations were similar across treatments.

In another percutaneous mouse study by Waechter and Rick (1988), 2000 mg/kg of ^{14}C-TEA was topically applied in water (200 μL, pH 7.0) to 21 male mice equipped with a glass ring, and the results were compared to the blood concentration–time curve following application of 2000 mg/kg TEA with no vehicle. The use of water as a vehicle resulted in blood concentrations similar to those produced by the application of neat material. Urine was the primary route of elimination following either intravenous or topical administration. Approximately 65% of the dose was eliminated within 24 hr. Feces was a secondary route for both routes of dosing, with 16.3%–27.7% being eliminated. The average amount of dose being retained in the body at sacrifice was 3.6% for the IV-administered animals, 4.5% and 7.8%, respectively, for the animals topically administered 1000 and 2000 mg/kg. The percutaneous studies of Waechter and Rick (1988) were used to calculate the absorption rate data given in Table 11. The rates of absorption are based on 24- and 48-hr experimental time periods. According to the data, TEA is almost completely absorbed in 24 hr, and use of 24 hr in the calculation gives equivalent rates across all studies of about 800 μg cm^{-2} hr^{-1} and k_p values of 3.6 $\times$ 10^{-2} cm hr^{-1}.

Waechter and Rick (1988) investigated the percutaneous absorption of 1000 mg/kg of TEA by three male rats. ^{14}C-TEA (200 μL) was applied to skin within a glass ring glued to the back of each animal. The ring was covered with a round glass cover slip to prevent access to the site by the animal while grooming. The animals were housed in all-glass Roth metabolism cages for the separate and complete collection of urine and feces at 6, 12, 24, and 48 hr post-administration. Blood samples were taken via an indwelling jugular catheter at 0.5, 1, 2, 4, 6, 24, and 36 hr

Table 11. *In vivo* Percutaneous absorption of ^{14}C-TEA in mice and rats.

	Single dose, 24- and 48-hr duration				
C3H/HeJ mice					
Applied dose, Unwashed					
mg/kg	1000[a]	2000[b]	2000[c]	2000[d]	2000[e]
µg/cm^3	20,400	23,314	23,314	20,400	23,314
Absorption					
µg cm^{-2} hr^{-1}	778	406	835	408	418
kp, cm/hr	3.8×10^{-2}	1.7×10^{-2}	3.6×10^{-2}	2.0×10^{-2}	1.8×10^{-2}
Rats F-344 (48-hr single dose)					
Applied dose, unwashed					
mg/kg	1000[f]				
µg/cm^3	132,000				
Absorption					
µg cm^{-2} hr^{-1}	2447.5				
kp, cm/hr	1.85×10^{-2}				

[a] Acetone/1.0 cm^2, 9 mice, probe study, sacrificed at 24 hr.
[b] Water, ring 1.5 cm in diameter (1.75 cm^2), 21 mice, probe study, sacrificed at 48 hr.
[c] Neat, 1.5-cm-diameter ring (1.75 cm^2), 3 mice, probe study, sacrificed at 24 hr.
[d] Neat, 2 cm^2 surface area, 24 mice, sacrificed at 48 hr.
[e] Neat, 1.5-cm-diameter ring (1.75 cm^2), 24 mice, sacrificed at 48 hr.
[f] Neat, 1.5-cm-diameter ring (1.75 cm^2), 3 rats, Roth cages, sacrificed at 48 hr.
Absorption rates calculated by authors from percutaneous absorption studies of Waechter and Rick (1988).

post-administration. Animals were sacrificed after 48 hr; the dose site was excised along with the remaining skin. The liver, kidneys, and carcass were collected along with the cage and ring washings. Washings, excreta, and tissues were analyzed for radioactivity. The concentration of ^{14}C-TEA equivalents in rat blood indicated that TEA was not absorbed as rapidly by the rat as it was by the mouse. The radioactivity in blood was nondetectable ($<2 \times$ background) at 0.5, 1, 2, 4, and 6 hr post-administration. According to Table 11, the rate of absorption through rat skin is about three- to sixfold greater (2400 vs. 400–800 μg cm^{-2} hr^{-1}) than through mouse skin. This three- to sixfold increase is the result of a sixfold increase in the concentration of TEA topically applied to rat skin. The applied concentrations gave equivalent k_p values of 1.85×10^{-2} cm/hr for the rat and mouse.

VI. Subchronic and Chronic Toxicity
A. General

A number of studies have been undertaken to characterize the potential dermal, oral, or inhalation toxicity of MEA, DEA, and TEA in experimental animals. Summaries of the designs and significant findings of these studies are given in Tables 12, 13, and 14. In general, the vapor pressure and irritant characteristics of a particular ethanolamine appear to have dictated the route of administration used to assess repeated-dose toxicity. Thus, the more volatile and irritating MEA has been evaluated primarily via inhalation of vapor, while the less volatile and irritating TEA has been evaluated primarily via oral and dermal routes.

In general, common target tissues in test species, regardless of the route of administration, have been the liver or kidneys. Treatment-related changes, both adaptive and pathological, in these tissues are consistent with the apparent metabolism or sequestration of these molecules in hepatic tissues and their excretion primarily via the urine. Skin effects noted in animals administered ethanolamines via dermal application or, in some cases, whole-body inhalation exposure, reflect the potential irritant properties of these strongly alkaline materials, especially at the high concentrations used in high dose testing. It is also important to note that dermal administration of varying dosages of ethanolamines in a number of studies has entailed the use of variable amounts of undiluted (neat) or very concentrated test materials. This practice increases the likelihood of causing pathological alterations of skin or alterations of skin as a barrier to absorption with subsequent systemic effects at relatively low dosages. As a result, effects observed may not reflect the toxicity of more dilute dosing solutions delivering the same total dosage to the skin of an animal.

A number of additional histopathological changes have also been reported in tissues of animals administered relatively high dosages of ethanolamines. The significance of these latter changes, however, is often difficult to ascertain due to confounding factors in affected animals, such as in-

Table 12. Summary of repeated dose and subchronic toxicity data for monoethanolamine.

Route	Duration	Test animals	Dosage	Potential effects	Reference
Inhalation	Various; high conc. ≤ 5 d, low conc. ≤ 30 d	Rats, dogs, cats, G. pigs, mice (strains unspecified)	Up to 36 ppm vapor, 7 hr/d; up to 793 mg/m³ aerosol, 7 hr/d	No specific toxicity reported except that guinea pigs displayed breathing difficulties at ≥ 260 mg/m³	Teron et al. (1957)
Inhalation	Variable: rats, 3.5–13 wk; G. pigs, 3.5 wk; dogs, 4–13 wk	CFW rats (M&F), Hartley G. pigs (M), beagle dogs (M)	Variable: rats, 5, 12, 66 ppm; G. pigs, 15, 75 ppm; dogs, 6, 12, 26, 102 ppm "continuous" exposure (23.5 hr/d, 7 d/wk)	High exposures: mortality, severe stress, breathing difficulties, behavior changes, skin irritation, histopathological changes (lungs, nasal mucosa, G. pigs only; liver, kidneys, G. pig and dogs only) of all species. All exposure levels: skin histopathology	Weeks et al. (1960)
Inhalation	Various; ≤ 6 mon	Rats (strain unspecified)	80–160 ppm, 5 hr/d	Exposed rats had decreased body weights, altered hematological parameters, altered urine chemistries, altered hippuric acid synthesis. Concluded liver and kidneys are target tissues	Timofievskaya (1962) as reviewed by Binks et al. (1992)
Oral (feed)	4 wk	Rats (strain unspecified)	160–2670 mg kg⁻¹ d⁻¹	≥ 1280 mg kg⁻¹ d⁻¹: mortality, kidney and liver histopathology (presumed); ≥ 640 mg kg⁻¹ d⁻¹: "altered" liver and kidney weight	Smyth et al. (1951)

Table 13. Summary of repeated dose and subchronic toxicity data for Diethanolamine.

Route	Duration	Test animals	Dosage	Potential effects	Reference
Inhalation	9 d	Rats (strain unspecified)	25 ppm, continuous exposure	Increased liver and kidney weight with altered serum chemistries	Hartung et al. (1970)
Inhalation	13 wk	Rats (strain unspecified)	6 ppm, "workday schedule"	Mortality, decreased growth rate, increased lung and kidney weight	Smyth et al. (1951)
Oral (drinking water)	2 wk	Rats, Fischer-344	0, 630, 1250, 5000, 10,000 ppm; (males est. 77–1016 mg kg^{-1} d^{-1}; females est. 79–1041 mg kg^{-1} d^{-1})	Males and females: ≥5000 ppm, mortality; ≥2500 ppm, kidney histopathology; ≥1250 ppm, decreased body weight, altered serum chemistries; most dosages, anemia, altered urine chemistries; Males: ≥1250 ppm, increased kidney weight; Females: all dosages, increased kidney weight	Hejtmancik et al. (1987a); NTP (1992)
Oral (drinking water)	2 wk	B6C3F1 mice	0, 630, 1250, 5000, 10,000 ppm; (males est. 110–1362 mg kg^{-1} d^{-1}; females est. 197–2169 mg kg^{-1} d^{-1}	Males and females: 10,000 ppm, severe dehydration; ≥2500 ppm, increased liver weight, decreased thymus weight, lymphoid tissue depletion; Males: ≥2500 ppm, liver histopathology	Hejtmancik et al. (1987b); NTP (1992)

(continued)

Table 13. (*Continued*)

Route	Duration	Test animals	Dosage	Potential effects	Reference
Oral (drinking water)	≤7 wk	Rats (strain unspecified)	4000 ppm	Mortality, "liver and kidney damage," normocytic anemia without bone marrow depletion	Hartung et al. (1970)
Oral (feed)	≤13 wk	Wistar rats	0, 5, 20, 90, 170, 350, 680 mg kg^{-1} d^{-1}	≥170 mg kg^{-1} d^{-1}; mortality, kidney, liver, small intestine and lung histopathology; ≥90 mg kg^{-1} d^{-1}; increased liver and kidney weight	Smyth et al. (1951)
Oral (drinking water)	13 wk	Fischer-344 rats	0, 160 (F), 320, 630, 1250, 2500, 5000 (M) (ppm); (Males est. 25–436 mg kg^{-1} d^{-1}; females est. 14–242 mg kg^{-1} d^{-1})	Males: 5000 ppm, mortality; ≥2500 ppm, kidney and CNS (gonads, hypospermia)[a] histopathology; ≥630 ppm, anemia; all dosages, decreased body weight; Females: 2500 ppm, severe dehydration; ≥1250 ppm, CNS (adrenal cortex)[a] histopathology; ≥320 ppm, anemia; all dosages, decreased body weight, increased kidney[b] and liver[b] weight, kidney histopathology	Melnick et al. (1994a); NTP (1992)
Oral (drinking water)	13 wk	B6C3F1 mice	0, 630, 1250, 2500, 5000, 10,000 ppm; (males est. 104–	Males and females: ≥2500 ppm, mortality, severe dehydration, heart and salivary	Melnick et al. (1994b); NTP (1992)

			1674 mg kg^{-1} d^{-1}; females est. 142–1154 mg kg^{-1} d^{-1})	gland histopathology; all dosages, increased liver weight, liver histopathology; Males: ≥2500 ppm, decreased body weight; ≥1250 ppm, increased kidney weight, kidney histopathology; Females: ≥2500 ppm, increased heart weight, heart histopathology; ≥1250 ppm, decreased body weight	Hejtmancik et al. (1987c); Melnick et al. (1988); NTP (1992)
Dermal (skin painting)	2 wk	Fischer-344 Rats	0, 125, 250, 500, 1000, 2000 mg kg^{-1} d^{-1}, 5 d/wk (ethanol vehicle or neat)	Males and females: ≥1000 mg kg^{-1} d^{-1}, mortality; ≥250 mg kg^{-1} d^{-1}, anemia; all dosages, skin irritation; Males: ≥1000 mg kg^{-1} d^{-1}, altered serum chemistry and urinalysis parameters; ≥500 mg kg^{-1} d^{-1}, increased kidney weight; Females: ≥125 mg kg^{-1} d^{-1}, increased kidney weight[b]; all dosages, altered serum chemistry and urinalysis parameters	Hejtmancik et al. (1987c); Melnick et al. (1988); NTP (1992)
Dermal (skin painting)	2 wk	B6C3F1 mice	0, 160, 320, 630, 1250, 2500 mg kg^{-1} d^{-1}, 5 d/wk (ethanol vehicle or neat)	Males and females: 2500 mg kg^{-1} d^{-1}, mortality; ≥1250 mg kg^{-1} d^{-1}, skin irritation; ≥630 mg kg^{-1} d^{-1}, decreased thymus weight;	Hejtmancik et al. (1987d); Melnick et al. (1988); NTP (1992)

(continued)

Table 13. (*Continued*)

Route	Duration	Test animals	Dosage	Potential effects	Reference
				≥ 320 mg kg^{-1} d^{-1}, increased liver weight; all dosages, skin acanthosis	
Dermal (skin painting)	13 wk	Fischer-344 Rats	0, 32, 63, 125, 250, 500 mg kg^{-1} d^{-1}, 5 d/wk (ethanol vehicle)	Males and females: 500 mg kg^{-1} d^{-1}, mortality, decreased body weight; ≥ 125 mg kg^{-1} d^{-1}, skin irritation; all dosages, increased liver[b] and kidney weight, kidney histopathology; Males: ≥ 250 mg kg^{-1} d^{-1}, CNS histopathology; Females: 500 mg kg^{-1} d^{-1}, CNS histopathology	Melnick et al. (1994a); NTP (1992)
Dermal (skin painting)	13 wk	B6C3F1 mice	0, 80, 160, 320, 630, 1250 mg kg^{-1} d^{-1}, 5 d/wk (ethanol vehicle)	Males and females: 1250 mg kg^{-1} d^{-1}, kidney, heart, and salivary gland histopathology; ≥ 160 mg kg^{-1} d^{-1}, liver histopathology; all dosages, skin irritation; Males: ≥ 160 mg kg^{-1} d^{-1}, increased liver weight; Females: all dosages, increased liver weight	Melnick et al. (1994b); NTP (1992)

[a]Change was attributed to direct effect of treatment, inanition, and dehydration-related weight loss, or a combination of these.
[b]Change observed was not dose related and may or may not have been treatment related.

Table 14. Summary of old repeated dose and subchronic toxicity data for Triethanolamine.

Route	Duration	Test animals	Dosage	Potential effects	Reference
Inhalation	2 wk	Fischer-344 rats	0, 125, 250, 500, 1000, 2000 mg/m³, 6 hr/d, 5 d/wk	Males and females: 2000 mg/m³, decreased body weight; ≥500 mg/m³, increased kidney[a] weight	Mosberg et al. (1985a)
Inhalation	2 wk	B6C3F1 mice	0, 125, 250, 500, 1000, 2000 mg/m³ 6 hr/d, 5 d/wk	Males: 2000 mg/m³, increased kidney weight; all dosages, increased RBCs and decreased lymphocytes[a], Females: 2000 mg/m³, decreased body weight; all dosages, decreased thymus[a] and heart[a] weight	Mosberg et al. (1985b)
Oral (drinking water)	2 wk	Fischer-344 Rats	0, 0.5, 1, 2, 4, 8%; (males, est. 660–4390 mg kg⁻¹ d⁻¹; females est. 680–5070 mg kg⁻¹ d⁻¹)	Males and females: ≥4%, decreased body weight (severe dehydration); Females: 2%, increased kidney weight	Hejtmancik et al. (1985a)
Oral (drinking water)	2 wk	B6C3F1 mice	0, 0.5, 1, 2, 4, 8% (Males est. 1000–8000 mg kg⁻¹ d⁻¹; females est. 1200–8900 mg kg⁻¹ d⁻¹)	Males and females: 8%, decreased body weight, dehydration, histologic changes in liver (likely glycogen depletion); ≥4%, decreased thymus weight	Hejtmancik et al. (1985b)

(continued)

Table 14. (*Continued*)

Route	Duration	Test animals	Dosage	Potential effects	Reference
Oral (diet)	4 wk	Rats (strain unspecified)	5–2610 mg kg^{-1} d^{-1}	$\geq$80 mg kg^{-1} d^{-1}, mortality, decreased growth rates, multi-organ histopathology	Smyth et al. (1951)
Oral (feed)	$\leq$24 wk	Rats (strain unspecified)	0, 200, 400, 800, 1600 mg kg^{-1} d^{-1}	All dosages, kidney and liver histopatholgoy	Kindsvatter (1940)
Oral (gavage)	$\leq$24 wk	Guinea pigs (strain unspecified)	0, 200, 400, 800, 1600 mg kg^{-1} d^{-1}, 5 d/wk	All dosages, kidney and liver histopatholgoy	Kindsvatter (1940)
Dermal (skin painting)	2 wk	Fischer-344 rats	0, 141, 281, 563, 1125, 2250 mg kg^{-1} d^{-1}, 5 d/wk (neat)	Males and females: 2250 mg kg^{-1} d^{-1}, decreased body weight; $\geq$563 mg kg^{-1} d^{-1}, skin irritation	Hejtmancik et al. (1985c); Melnick et al. (1988)
Dermal (skin painting)	2 wk	B6C3F1 mice	0, 214, 427, 843, 1686, 3370 mg kg^{-1} d^{-1}, 5 d/wk (neat)	Males: all dosages, skin irritation; Females: $\geq$427 mg kg^{-1} d^{-1}, skin irritation	Hejtmancik et al. (1985d); Melnick et al. (1988)
Dermal (skin painting)	13 wk	Fischer-344 rats	0, 125, 250, 500, 1000, 2000 mg kg^{-1} d^{-1}, 5 d/wk (ethanol vehicle or neat)	Males and females: 2000 mg kg^{-1} d^{-1}, decreased body weight; altered serum chemistry, urinalysis, and hematological parameters; pituitary gland hypertrophy;	Hejtmanacik et al. (1987e); NTP (1994)

Route	Duration	Species	Dosages	Effects	Reference
Dermal (skin painting)	13 wk	B6C3F1 mice	0, 250, 500, 1000, 2000, 4000 mg kg^{-1} d^{-1}, 5 d/wk (ethanol vehicle or neat)	Males: ≥250 mg kg^{-1} d^{-1}, skin irritation; ≥500 mg kg^{-1} d^{-1}, increased kidney weight Females: ≥1000 mg kg^{-1} d^{-1}, increased kidney weight and pathology; ≥500 mg kg^{-1} d^{-1}, skin irritation	Hejtmancik et al. (1987f); NTP (1994)
Dermal (skin painting)	13 wk	C3H/HeJ mice	0, 160, 540, 2300 mg kg^{-1} d^{-1}, 3 d/wk (acetone vehicle)	Males and females: 4000 mg kg^{-1} d^{-1}, skin irritation, increased liver and kidney[a] weight; Males: all dosages, decreased body weight[a]; Females: 4000 mg kg^{-1} d^{-1}, increased spleen weight Males and females: all dosages, skin irritation	DePass et al. (1995)

[a]Change observed was not dose related and may or may not have been treatment related.

creased mortality, severe body weight depression, and stress-related changes. Obviously, the appropriateness of such data for risk assessment purposes must be weighed carefully. Finally, data have also been generated on the potential oncogenicity of a single ethanolamine, TEA, under a variety of study designs. The results of these bioassays have been somewhat contradictory and occasionally controversial but in general have failed to demonstrate clear evidence of a carcinogenic response for this compound. The results of MEA, DEA, and TEA toxicity studies ranging in duration from several weeks to several years are reviewed later, arranged by compound and route of exposure.

B. Monoethanolamine

Toxicity studies on MEA have been conducted in laboratory animals using inhalation and oral routes of administration (Table 12). As noted, the paucity of repeated-dose dermal toxicity data for this compound appears to reflect the relatively severe irritant properties of concentrated MEA solutions, which make repeated administration of high concentrations of MEA impossible. Indeed, the skin was noted in one study to be a significant target tissue of relatively high concentrations of MEA vapor (see following). Apart from portal-of-entry tissues, the kidneys and livers of test species were observed to be the most sensitive systemic target tissues of MEA. Several other tissues have also been reported to be affected by treatment, but only at relatively high dosages at which interpretation of results was often confounded by test material rejection, severe stress, loss of body weight, and a variety of secondary effects.

1. Subchronic Toxicity

Inhalation.   The potential subchronic inhalation toxicity of MEA has been evaluated in several early studies using a number of species and several exposure regimens. Treon et al. (1957) exposed dogs, cats, guinea pigs, rats, and mice to relatively high concentrations of MEA vapor and aerosol under a number of exposure regimens, including the exposure of animals to approximately 793 mg/m^3 MEA (primarily as an aerosol), 7 hr/d for 5 d or to approximately 126 mg/m^3 MEA (primarily as a 36-ppm vapor), 7 hr/d for 25 of 30 d. No mortality or other signs of toxicity were noted; however, guinea pigs "exhibited gasping" or "breathed at an unduly rapid rate" when exposed to concentrations of MEA ≥ 660 and ≥ 260 mg/m^3, respectively, during exposure. Upon postmortem examination, exposed animals were reported to have "normal viscera."

In a subsequent study, Weeks et al. (1960) exposed dogs, rats, and guinea pigs to varying concentrations of a highly purified MEA vapor using essentially continuous exposure conditions (23.5 hr/d), 7 d/wk, for various time periods. Male or female CFW rats were exposed to 5 ppm (12 mg/m^3)

vapor for 40 d, 12 ppm (30 mg/m^3) for 90 d, or 66 ppm (164 mg/m^3) for 24 d. Male Hartley guinea pigs were exposed to 15 ppm (37 mg/m^3) or 75 ppm (187 mg/m^3) for 24 d. Male beagle dogs were exposed to 6 ppm (15 mg/m^3) vapor for 60 d, 12 ppm (30 mg/m^3) or 26 ppm (65 mg/m^3) for 90 d, or 102 ppm (254 mg/m^3) for 30 d. A significant amount of dermal exposure to MEA occurred at all exposure levels as MEA vapor reportedly condensed onto the surface of exposed animals, causing their pelts to become "wet, matted, and greasy" at higher exposure concentrations and discolored or "slightly greasy" at lower concentrations.

All high dose group animals in the Weeks et al. (1960) study displayed pronounced clinical signs of skin and respiratory irritation, which progressed with time to hair loss, severe skin lesions (e.g., ulceration), moist rales and fever in dogs, and breathing difficulties in rats and guinea pigs. The authors suggested that contact with wetted caging surfaces exacerbated skin lesions. Body weight data were not reported, but it was noted that feed consumption in dogs was decreased. Mortality as high as 80% occurred in high dose groups of all three test species (37/45 rats at 66 ppm, 17/22 guinea pigs at 75 ppm, 1/3 dogs at 102 ppm MEA).

In general, behavior changes reported by Weeks et al. (1960) for exposed animals appear to have reflected the extreme irritancy of the MEA atmospheres employed. Signs of severe discomfort and irritancy progressed to pronounced apathy, lethargy, and depression. In dogs, leg muscle tremors were also noted; however, the toxicological significance of this finding is questionable, as the feet of these animals were ulcerated and "became so sensitive they would not voluntarily walk." At exposure concentrations of 12–26 ppm vapor, similar though less severe effects were observed in the respective species of animals. At exposure levels of 5–6 ppm vapor, dogs and rats had discolored coats, alopecia, and skin irritation, and appeared less active than controls.

Microscopic examination of tissues excised from dogs, rats, and guinea pigs inhaling MEA vapor identified the skin, lungs, nasal mucosa, liver, and kidneys as potential target tissues (Weeks et al. 1960). Histopathological changes in the skin of all species of high dose group animals (exposure to 66–102 ppm MEA) included vacuolation of epithelial cells, thickening of the epithelium, inflammation, and necrosis. In more severe cases, the skin was essentially eliminated as a barrier to absorption of MEA because necrosis extended to the level of the underlying muscle. In contrast, the epithelium of the skin of animals exposed to 5–15 ppm MEA was only slightly increased in thickness and had an increased number of exfoliated cells relative to the skin of control dogs.

Histopathological changes in other tissues were observed only in high dose group animals. Respiratory tract changes consisted of lymphocytic infiltrate in the lungs of rats and guinea pigs and a generalized congestion and focal hemorrhages in dogs. In addition, the nasal epithelium of the latter animals showed slight erosion and plasma cell infiltration. The liver parenchyma of one or both rodent species and dogs displayed areas of cloudy cytoplasm or hypertrophy. Central vein congestion and hepatocellular vacuolation was also observed in dogs, while a panlobular "fatty metamorphosis" was also noted in the livers of rats and guinea pigs. Finally, the renal tubular epithelium of dogs was observed to have increased "hyaline pink granules" and "cloudy swelling" relative to controls. The latter change was also noted in the kidneys of high dose group guinea pigs. No treatment-related changes were observed in any tissues other than skin in animals exposed to lower concentrations of MEA vapor.

Finally, Timofievskaya (1962) reported on a Soviet study in which rats were exposed to concentrations of technical-grade MEA (75% purity) ranging from 80 to 160 ppm (200–400 mg/m^3), 5 hr/d, for up to 6 mon. As reviewed by Binks et al. (1992), exposed animals suffered decreased body weight, changes in several hematological parameters, diuresis, proteinuria, and reduced synthesis of hippuric acid. The authors concluded that liver and kidneys were target tissues for inhaled MEA in rats and did not specify an exposure level lacking any effects.

Oral.   Limited data in dogs and rats are available on the repeated-dose oral toxicity of MEA. Smyth et al. (1951), as part of a series of range-finding studies, provided rats with feed formulated to deliver dosages of MEA (purity unknown) ranging from 160 to 2670 mg kg^{-1} d^{-1} for 30 d (complete dosage information was not provided). The authors reported "altered" liver or kidney weight(s) in rats ingesting ≥ 640 mg kg^{-1} d^{-1} MEA and unspecified "microscopic lesions" (likely in liver or kidney) and death at dosages ≥ 1280 mg kg^{-1} d^{-1}.

The oral toxicity of MEA following chronic ingestion has also been indirectly evaluated in a dog bioassay of a "composite of dyes and base components found in semipermanent hair color products" (Wernick et al. 1975). In this relatively comprehensive study, male and female beagle dogs were administered the composit formulation via their feed at dosages of 0, 19.5, or 97.5 mg kg^{-1} d^{-1} for up to 2 yr. The corresponding dosages of MEA were 0, 4, and 22 mg kg^{-1} d^{-1}. No treatment-related effects were observed in a number of parameters evaluated, including body weight, hematological and clinical chemistry parameters, urinalysis parameters, clinical signs, organ weight, and gross and microscopic pathology.

C. Diethanolamine

Toxicity studies on DEA have been conducted in rats and mice using inhalation, oral, and dermal routes of administration. Consistent with other ethanolamines, the skin, kidneys, or liver of test species are observed to be the most sensitive target tissues of DEA. A number of other tissues also appear to be affected by treatment but only at relatively high dosages, where interpretation was often confounded by test material rejection, stress, loss of body weight, and secondary effects. In addition, DEA also causes a rat-specific microcytic anemia that does not involve the bone marrow and appears to be unique among this family of compounds. Data suggest that rats, are somewhat more sensitive than mice to the toxic effects of DEA. In rats, a gender difference also occurs, with females generally being more sensitive than males. In mice, this distinction is not as clear; if anything, female mice appear somewhat more resistant than male mice.

It is important to note that a potentially confounding factor in any toxicological evaluation of DEA, and many other secondary amines, is the potential formation of genotoxic nitrosamines in the presence of a nitrosating compound either in test diets, in drinking water, or *in situ* in gastric fluid. Indeed, the formation of the nitrosamine *N*-nitrosodiethanolamine (NDELA) has been observed to occur in mice administered DEA dermally while imbibing drinking water containing a relatively high level of sodium nitrite, a potential contaminant of water supplies (Preussmann et al. 1981). Oral dosages of NDELA as low as 1-2 mg kg^{-1} d^{-1} have been reported to cause increases in liver tumors in rats (Lijinsky and Kovatch 1985; Preussmann et al. 1982; TRGS 1988).

1. Subchronic Toxicity

Inhalation.   Limited inhalation toxicity data have been generated on DEA (purity unspecified). Hartung et al. (1970) reported a study in which rats were exposed to relatively high concentrations of DEA vapor (200 ppm) or aerosols (1400 ppm). While there is a paucity of experimental detail available, exposed animals apparently experienced respiratory difficulties and mortality. Inhalation of 25 ppm DEA continuously (23.5 hr/d) for 9 d reportedly caused increased liver and kidney weights, which were accompanied by elevated serum aspartate aminotransferase activity and BUN levels. Inhalation of 6-ppm vapor by male rats on a "workday schedule" for 13 wk reportedly caused depressed growth rates, increased lung and kidney weights, and mortality.

Oral.   The oral toxicity of DEA has been comprehensively examined in rats and mice. Orally administered DEA has also been implicated in the poisoning of dogs and cats and has been evaluated in humans as a possible therapeutic agent.

In an early study, Smyth et al. (1951) administered male Wistar rats 0, 5, 20, 90, 170, 350, or 680 mg kg^{-1} d^{-1} DEA (purity unknown) via their feed for up to 13 wk. Increased mortality was noted in rats ingesting $\geq$ 170 mg kg^{-1} d^{-1}, while the liver and kidney weights of rats ingesting $\geq$ 90 mg kg^{-1} d^{-1} DEA were increased relative to controls. Histopathological changes consisting of a cloudy swelling and degeneration of renal tubular epithelial cells and fatty degeneration of the liver occurred in rats ingesting $\geq$ 170 mg kg^{-1} d^{-1} DEA (see CIR 1983). Additional changes occurred in tissues of the small intestines and lungs of these latter rats. No treatment-related effects were observed in rats ingesting ≤ 20 mg kg^{-1} d^{-1} DEA.

In a relatively abbreviated study, Hartung et al. (1970) reported administering DEA to rats as a neutralized solution in drinking water at a concentration of 4000 ppm for up to 7 wk. Little experimental detail is available, but imbibed DEA was reported to cause mortality and "liver and kidney damage." In addition, a pronounced normocytic anemia without bone marrow depletion or increase in the number of reticulocytes was also reported.

More recently, several comprehensive oral toxicity studies have been conducted as part of a National Toxicology Program (NTP) examination of DEA toxicity in rats and mice. In an initial repeated-dose study, male and female Fischer-344 rats were provided drinking water (neutralized with HCl) containing 0, 630, 1250, 2500, 5000, or 10,000 ppm of highly purified DEA for up to 2 wk (Hejtmancik et al. 1987a; NTP 1992). Mean calculated dosages administered to both sexes of rats ranged from approximately 77 to 1041 mg kg^{-1} d^{-1} DEA. Numerous treatment-related effects were observed as well as those secondary to severe dehydration and starvation. Antemortem changes included clinical signs of toxicity and mortality in high dose group males and in females imbibing water containing ≥ 5000 ppm DEA; depressed water consumption and body weight of all treated rats; decreased serum proteins and elevated blood urea nitrogen (BUN) in all treated males and females imbibing ≥ 1250-ppm solutions; and anemia, enzymuria, increased urinary nitrogen, and increased urine specific gravity in most treatment groups of males and females. Postmortem changes included a number of changes in organ weights that appeared secondary to decreased body weight of treated rats, and treatment-related increases in kidney weight of males imbibing ≥ 1250-ppm solution and all dose groups of females. Renal tubule necrosis of the "chronic type" accompanied kidney weight changes in both males and females imbibing drinking water containing ≥ 2500 ppm DEA.

In a subsequent subchronic toxicity study, male and female Fischer-344 rats were provided drinking water (neutralized with HCl) containing 0, 160 (females only), 320, 630, 1250, 2500, or 5000 (males only) ppm of highly purified DEA for up to 13 wk (Melnick et al. 1994a; NTP 1992). Mean calculated dosages ranged from 25 to 436 mg kg^{-1} d^{-1} DEA in males and from 14 to 242 mg kg^{-1} d^{-1} DEA in females. Decreases in water consumption and clinical signs of toxicity, including emaciation, roughened coat, and tremors, characterized animals in the highest dose levels. Two of 10 high dose group males died prior to scheduled necropsy. The terminal body weights of all groups of treated male and female rats were depressed in a dose-related manner from 7% to 44% and from 5% to 26%, respectively, relative to controls. In males, weight changes in liver, testes, epididymus, and kidneys, with the possible exception of high dose group animals, reflected changes in body weight. In females, kidney and liver weights appeared elevated 26%–36% and 15%–28%, respectively, but not in a dose-related manner.

Histologically, the most sensitive treatment-related effects of DEA were observed in the kidneys of treated females. This lesion, consisting primarily of "tubules lined by epithelial cells with more basophilic staining of the cytoplasm and a higher nuclear/cytoplasmic ratio" and "tubular mineralization," occurred in control and treated animals alike. However, the severity or incidence of the lesion increased in a dose-related manner in females imbibing the DEA solutions of 160–1250 ppm or greater. These changes were interpreted to reflect a regenerative change as supported by the occurrence of tubular cell necrosis in males and females imbibing ≥ 2500-ppm and ≥ 1250-ppm DEA solutions, respectively. Interestingly, the latter effect was limited to similar dosages in rats imbibing DEA for only 2 wk, suggesting a lack of progression of this lesion over time. In addition to renal changes, these same, relatively toxic, dosages of DEA were also associated with demyelination of nerve tracts in medullary brain and spinal cord tissues. It was suggested that the latter effect possibly resulted from the misincorporation of DEA into cell membrane lipids.

As noted earlier by Hartung et al. (1970), imbibed DEA caused a moderate, microcytic, normochromic anemia in both sexes of rats (Melnick et al. 1994a; NTP 1992). Dose-related decreases in erythrocyte counts (7%–35%), hemoglobin concentrations (10%–40%), and hematocrit (9%–34%) were observed in males and females imbibing ≥ 630- and ≥ 320-ppm DEA solutions, respectively. A slight, yet statistically identified, decrease in hemoglobin concentration (3%) also occurred in low dose group males. The microcytic nature of this effect was evidenced by parallel decreases in the mean corpuscular volume of these rats. Consistent with the results of earlier studies, hematological changes were not accompanied by histological alterations in bone marrow, indicating that development of anemia was not a result of damage to hematopoetic tissues. Rather, hematological effects appear to reflect the incorporation of DEA into red blood cell membranes (Mathews and Jeffcoat 1991; Mathews et al. 1993; Waechter et al. 1995). Overall, a NOEL was not established for rats imbibing DEA. However, given the relatively minor hematological changes in males, the minimal severity of renal effects in females, and the lack of dose–response in kidney weight changes in females, the lowest dosages appear to be at or close to the no-observable-adverse-effect level (NOAEL) for imbibed DEA.

Repeated-dose and subchronic oral toxicity studies of DEA have also been conducted in mice. Male and female $B_6C_3F_1$ mice were provided drinking water (neutralized with HCl) containing 0, 630, 1250, 2500, 5000, or 10,000 ppm of high purity DEA for up to 2 wk as part of the NTP evaluation of DEA (Hejtmancik et al. 1987b; NTP 1992). Mean calculated dosages administered to males ranged from 110 to 1362 mg kg^{-1} d^{-1} and those to females from 197 to 2169 mg kg^{-1} d^{-1} DEA. The 10,000-ppm DEA solution appeared to be unpalatable to both sexes of mice; however, despite signs of severe dehydration in these animals, no mortality was observed. In contrast to findings in the rat, the liver rather than the kidneys was the primary target tissue of imbibed DEA in mice. Liver weight was elevated in both sexes of mice drinking a ≥ 2500-ppm DEA solution relative to controls. Histopathological changes consisting of "cytoplasmic vacuolization and degeneration of (centrilobular) hepatocytes" accompanied liver weight changes in males. In addition, both sexes of mice imbibing a ≥ 2500-ppm DEA solution had lymphoid depletion and decreased thymus weight, suggesting severe stress. No significant treatment-related effects were observed in mice imbibing a 630-ppm DEA drinking water solution.

In a subsequent mouse subchronic toxicity study, male and female $B_6C_3F_1$ mice were provided drinking water containing 0, 630, 1250, 2500, 5000 or 10,000 ppm of DEA for up to 13 wk (Melnick et al. 1994b; NTP 1992). Mean calculated dosages administered ranged from 104 to 1674 mg $kg^{-1} d^{-1}$ DEA in males and from 142 to 1154 mg $kg^{-1} d^{-1}$ DEA in females. All males and females of the two highest dose groups and three females of the 2500-ppm dose group died prior to study termination. Decreases in water consumption and clinical signs of toxicity, including emaciation, roughened coat, and tremors, characterized survivors in the 2500-ppm dose groups. The body weights of mice were depressed roughly 10% in males imbibing 2500 ppm DEA and approximately 9% and 18% in females imbibing 1250 and 2500 ppm DEA, respectively. As in the repeated-dose study, the most sensitive target tissue identified in mice was the liver. Liver weights were increased in a dose-related manner in all surviving dose groups of males and females from 13% to 29% and from 28% to 85%, respectively.

Weight changes were accompanied by increases in serum alanine aminotransferase and sorbitol dehydrogenase activities and histopathological changes. Nearly all treatment groups had cellular "hypertrophy with increased eosinophilia and disruption of hepatic cords," increased incidence of "nuclear pleomorphism," and the "frequent presence of large, multinucleated hepatocytes." No clear NOEL was identified in either sex for this effect. The livers of both sexes of mice imbibing ≥ 2500 ppm solutions of DEA also had individual cell necrosis or foci of necrotic hepatocytes.

A number of additional target tissues of DEA were identified at relatively high oral dosages in mice (Melnick et al. 1994b; NTP 1992). Kidney weights of males imbibing 1250 or 2500 ppm DEA solutions were elevated approximately 10% and 14%, respectively, relative to controls. Histologically, renal tissues from these latter animals had a lesion similar to that observed in the kidneys of rats; slightly smaller, more basophilic, tubule epithelial cells. Minimally increased heart weight was accompanied by degenerative histopathological changes in nearly all high dose group females and a single high dose group male. Finally, histopathological changes consisting of a decrease in secretory granules and concomitant hypertrophy of secretory acini were observed in the submandibular salivary glands of male and female mice imbibing ≥ 2500 ppm DEA. Similar changes were not observed in other salivary gland tissues.

DEA has also been implicated in a number of poisoning cases of dogs and cats in South Florida related to the repeated ingestion of an over-the-counter flea treatment (Sundlof and Mayhew 1983). The product in question contained approximately 53% DEA in an aqueous solution that was added to the animals' feed at a recommended dosage of approximately 44 mg/kg. The exact dosages actually administered the 39 dogs and 12 cats affected, however, were not reported. The onset of neurological symptoms, primarily tremors, stiffness, and ataxia progressing to paresis and paralysis, was reported by pet owners to occur from 2 d to 18 mon following the start of dosing. A number of other symptoms were also reported, including emesis, diarrhea, convulsions, bleeding, anemia, and elevated serum BUN levels and alanine aminotransferase activity. A mortality rate of 41% was reported in both species; however, in less severely affected animals, symptoms resolved upon cessation of treatment.

Finally, DEA has been examined as a potential therapeutic agent for persons suffering diabetes mellitus, thrombophlebitis, and hyperlipidemia. Six newly diagnosed adult-onset diabetics were dosed orally with 250 mg DEA salt for 3–140 d (Roehm and Hooberry 1959). The authors reported that "each patient's tolerance (glucose) was more nearly normal." DEA was evaluated for its lipid-lowering potential in subjects administered dosages of 250 mg and 500 mg or 1000 mg for 10–18 wk and 30 d, respectively. No significant treatment-related changes were observed with the exception of a reported general lowering of blood lipids, including cholesterol, and a possible increased clotting time. All other hematological, clinical chemistry or urine chemistry parameters, and liver function tests were observed to be normal. Subsequent administration of 500 mg DEA to 33 patients twice daily for 6 mon also resulted in a lowering of blood lipids (Roehm 1973, 1975). Finally, DEA was observed to decrease fibrinogen levels and increase blood clotting times in several thrombophlebitis patients administered 500 mg/d of DEA for at least a week (Roehm 1973). Based on the latter two studies, it was suggested that DEA may have a direct anticoagulant effect in humans (Roehm 1973).

Dermal. The potential toxicity of DEA following dermal exposure has been examined in several rat and mouse skin painting studies. As noted, the fact that an occluding wrap was not employed in these studies to cover the application site suggests that an undetermined amount of the administered dose was ingested by test animals during grooming.

Several fairly comprehensive dermal studies have also been conducted as part of the NTP examination of DEA toxicity in rats and mice. In an initial repeated-dose study, dosages of 0, 125, 250, 500, 1000, or 2000 mg kg^{-1} d^{-1} of highly purified DEA as an ethanol solution or neat were applied to the clipped backs of male and female Fischer-344 rats 12 times over a 16-d interval (Hejtmancik et al. 1987c; Melnick et al. 1988). A number of direct treatment-related and secondary (e.g., stress-related) effects were observed. Antemortem changes included irritation of skin at the application site in both sexes of rats at ≥ 500 mg kg^{-1} d^{-1}; clinical signs of severe toxicity or mortality in both sexes of rats administered ≥ 1000 mg kg^{-1} d^{-1} DEA; decreased serum proteins and elevated BUN in males administered ≥ 1000 mg kg^{-1} d^{-1} and females administered ≥ 500 mg kg^{-1} d^{-1} DEA; a microcytic anemia in all rats administered ≥ 250 mg kg^{-1} d^{-1} DEA; and enzymuria and increased urinary nitrogen in all treatment groups of females.

Postmortem, a number of organ weight changes were observed. Kidney weights were elevated relative to controls in most dose groups of females and the higher dose groups of males. Histopathological changes were observed only in the skin and kidneys of treated rats. These included ulcerations, chronic inflammation, and acanthosis of application site skin of males and females at all dosages, renal tubular necrosis in both sexes of high dose group rats, and increased mineralization of some tubules in female rats administered ≥ 500 mg kg^{-1} d^{-1} DEA. Overall, apart from altered kidney weights in females that were not dose related, there was little suggestion of systemic toxicity and only relatively minor skin effects at the 125 mg kg^{-1} d^{-1} dosage level.

In a subsequent subchronic study, dosages of 0, 32, 63, 125, 250, or 500 mg kg^{-1} d^{-1} of DEA in a 95% ethanol vehicle were applied to the shaved backs of male and female Fischer-344 rats 5 d/wk for 13 wk (Melnick et al. 1994a; NTP 1992). Dermal irritation and crusting were evident at the site of DEA application in both sexes of rats administered ≥ 125 mg kg^{-1} d^{-1}, and some mortality was observed in high dose group rats. Systemic effects caused by dermally administered DEA were similar to those caused by orally administered DEA. These included body, liver, and kidney weight changes; histopathological changes in kidney and brain tissues; and anemia (in the absence of bone marrow pathology).

The body weights of males and females administered relatively high dermal dosages of DEA in the NTP study were decreased as much as 30% relative to controls. The liver and kidney weights of nearly all groups of treated rats were increased; however, these changes generally lacked dose-response. A possible exception to this was the dose-related increase in liver weight of treated females. Systemic histopathological changes, however, only accompanied kidney weight changes. Nephropathy, evidenced primarily as changes in tinctorial properties and the ratio of nuclear to cytoplasmic areas similar to that observed in the DEA oral toxicity studies, was reported to occur in all dose groups of females, including controls. These changes, however, lacked dose–response in both incidence and severity. In males, the only kidney pathological change observed was an increase in tubule mineralization at the highest dosage. Frank tubular necrosis was observed only in females at dosages ≥ 250 mg kg^{-1} d^{-1}.

A number of other changes attributed to dermal administration of DEA were noted in the skin, brain, and blood application sites of treated rats (Melnick et al. 1994a; NTP 1992). Ulcers and chronic inflammation were evident in treated skin at higher dosages, while acanthosis and hyperkeratosis occurred in males administered ≥ 63 mg kg^{-1} d^{-1} and all dose groups of females. Demyelination of nerves within the medulla oblongata of the brain occurred in one or both sexes of rats administered the highest dosage(s) of DEA (250 or 500 mg kg $^{-1}$ d $^{-1}$). The authors reported that this latter lesion, while morphologically similar to that observed in rats imbibing relatively high dosages of DEA, was generally less severe and lacked spinal cord involvement. Finally, consistent with findings for orally administered DEA in rats, the erythrocyte count, hemoglobin concentration, hematocrit, and mean corpuscular volume were decreased in a dose-related manner in all dose groups of rats with the exception of low dose group males. Overall, given the minimal degree of treatment-related changes in any tissue of this latter group, a dosage of 32 mg kg $^{-1}$ d $^{-1}$ would appear to represent a NOAEL for dermally administered DEA in male rats. The increased incidence of nephropathy, tubule mineralization, and skin histopathological changes that accompanied organ weight changes in low dose group females, however, suggests a more significant effect of dermally administered DEA in female rats and lack of a clear NOAEL.

Repeated-dose and subchronic dermal toxicity studies of DEA have also been conducted in mice as part of the NTP evaluation of DEA. Dosages of 0, 160, 320, 630, 1250, or 2500 mg kg $^{-1}$ d $^{-1}$ of highly purified DEA as an ethanol solution or neat were applied to the clipped backs of $B_6C_3F_1$ mice 12 times over a 16-d interval (Hejtmancik et al. 1987d; Melnick et al. 1988; NTP 1992). Similar to rats, nearly all mice repeatedly administered 2500 mg kg $^{-1}$ d $^{-1}$ DEA failed to survive the dosing period. Signs of severe toxicity (e.g., emaciation, abnormal posture, hypoactivity) were noted in these and the mice dosed with 1250 mg kg $^{-1}$ d $^{-1}$. Grossly, the skin at the site of application in these animals was reported to be irritated and ulcerated. Histological evaluation revealed ulcers and chronic inflammation. Acanthosis occurred in the skin of all treatment groups of mice. Systemically, liver weights of both sexes of mice were elevated relative to controls at dosages of ≥ 320 mg kg $^{-1}$ d $^{-1}$; however, no histopathological changes accompanied this weight change. Thymus weights were also decreased in males and females dosed with ≥ 160 and ≥ 630 mg kg $^{-1}$ d $^{-1}$ DEA, respectively, suggesting stress. No other significant treatment-related changes were noted. Overall, apart from acanthosis of the skin at the application site, no treatment-related effects were evident in mice dermally administered 160 mg kg $^{-1}$ d $^{-1}$ DEA for 2 wk.

In a subsequent subchronic toxicity study, dosages of 0, 80, 160, 320, 630, or 1250 mg kg $^{-1}$ d $^{-1}$ DEA in a 95% ethanol vehicle were applied to the shaved backs of male and female $B_6C_3F_1$ mice 5 d/wk for 13 wk (Melnick et al. 1994b; NTP 1992). Dermal irritation and crusting were evident at the site of DEA application in both sexes of mice administered ≥ 630 mg kg $^{-1}$ d $^{-1}$. Several high dose males and females failed to survive to the end of the study; however, body weights of treated mice appeared to be relatively unaffected by administration of DEA. Histologically, acanthosis was observed in the skin of all treated mice, while at higher dosages, ulcers, chronic inflammation, and hyperkeratosis were also noted.

Target tissues of dermally administered DEA distant from the site of application were generally similar to those found in mice imbibing DEA. A pronounced dose-related increase in liver weights up to 48% and 92% was noted in males administered ≥ 160 mg kg^{-1} d^{-1} DEA and in all dose levels of females, respectively. In general, histopathological effects in the livers of these mice consisted of altered tinctorial characteristics, disruption of the lobular pattern, and increased nuclear pleomorphism. At relatively high dosages, multinucleated giant hepatocytes and (in males) small foci of co-agulative necrosis were also observed. In addition to hepatic effects, dose-related increases in kidney weight were noted in both sexes of treated mice; however, unlike the effects in liver, other organ weight changes were not accompanied by histopathological changes. Necrosis of kidney tubules was noted in only about one-half of the males and only a single female adminis-tered a potentially lethal dosage of DEA, 1250 mg kg^{-1} d^{-1}. Other treat-ment-related effects observed at this relatively high dosage included increased heart weight, degenerative changes in cardiac tissues, and hyper-trophy of acini of the submaxillary salivary glands. Overall, the occurrence of organ weight changes and liver and skin histopathology in low dose group males suggests that a NOAEL was not identified. In contrast, the lack of systemic histopathological changes in females suggests that 80 mg kg^{-1} d^{-1} represents an NOAEL under the conditions of the study for these mice.

2. Chronic Toxicity and Oncogenicity. A 2-yr DEA chronic dermal toxic-ity study was conducted at Battelle-Columbus Laboratories for NTP (Re-search Triangle Park, NC) using B$_6$C$_3$F$_1$ mice. Dosages of 0, 40, 80, or 160 mg kg^{-1} d^{-1} were topically applied to back skin. At the time of this review, the results of the pathological examination have not been published by NTP. DEA was negative in the TG.AC transgenic mouse bioassay at a dermal dose of 20 mg/mouse/d, 5d/wk (Tennant et al., 1995).

D. Triethanolamine

TEA toxicity studies of varying duration have been conducted in rats and mice utilizing inhalation, oral, and dermal routes of administration. The design and significant findings of these studies are outlined in Table 14. In addition, five chronic toxicity/oncogenicity studies of varying quality have also been conducted in which TEA has been administered via the oral or dermal route of exposure. A striking feature of all of these studies has been the relatively high dosages tolerated by test animals, especially mice, for prolonged periods of time. As with other ethanolamines, the kidneys and livers of test species have typically been observed to be the most sensitive systemic target tissues of TEA, while the skin represents a target tissue in dermal toxicity studies. Consistent with the lack of genotoxicity of TEA in standard assays, results of oncogenicity bioassays have generally been negative. As noted, there have been a few reports of relative increases in specific tumor types in animals chronically administered high dosages of TEA, but in each case a number of potentially confounding factors have left questions regarding the accuracy of the findings or their relevance in terms of human risk assessment.

One potential confounding study result, that of nitrosamine formation in test diets, drinking water, or *in situ* in gastric fluids, would not appear to be as significant in the toxicological evaluation of TEA as for DEA studies. An initial report by Lijinsky et al. (1972) suggested that TEA may undergo nitrosative dealkylation and form NDELA under mildly acidic conditions. This conclusion was based, however, upon the reaction of molar concentrations of reactants over a 16-hr period. A subsequent study by Mervish (1975), as reviewed by Douglas et al. (1978), demonstrated only negligible potential for TEA to form NDELA under physiological conditions, including gastric pH. It was noted that nitrosation of tertiary amines proceeded at a rate roughly 10,000 times slower than the nitrosation of a secondary amine, even under acidic conditions. Only in the presence of significant amounts of a DEA impurity and nitrosating agent in feed or drinking water would significant levels of NDELA formation be likely to occur during the bioassay of TEA. However, it is also important to note that the analytical determination of NDELA in dosing solutions or diets, and thus the potential formation of this nitrosamine *in situ*, have not typically been carried out.

1. Subchronic Toxicity

Inhalation. The relatively low vapor pressure of TEA dictates that any evaluation of TEA inhalation toxicity be conducted using respirable aerosols. Several such studies, albeit of relatively short duration, have been conducted in both rats and mice as part of an NTP-sponsored evaluation of DEA toxicity to rats and mice. In the first of these, male and female Fischer-344 rats were exposed to 0, 125, 250, 500, 1000, or 2000 mg/m^3 TEA aerosol for 6 hr/d, 5 d/wk, over a 16-d period (Mosberg et al. 1985a). Both sexes of rats exposed to 2000 mg/m^3 had depressed body weight gains relative to controls, resulting in roughly 7%–8% lower terminal body weights. In addition, the kidney weights of males and females exposed to ≥ 500 mg/m^3 TEA aerosol were noted to be elevated; however, these changes generally lacked dose–response. Furthermore, increased kidney weights were not associated with any gross or histopathological change.

Indeed, the only histological change reported in exposed rats that was potentially treatment related was a "minimal to slight acute inflammation" of the submucosa of the larynx. The significance of this lesion, however, was questioned by the authors due to its lack of dose–response and its sporadic occurrence within the sections of larynx examined. In addition, no lesions were reported in the nasal mucosa or pulmonary tissues of exposed rats, suggesting a lack of severe irritation from TEA exposure. Finally, in contrast to DEA effects in rats, no treatment-related changes were observed in hematological parameters measured. It may be concluded that exposure to 250 mg/m^3 TEA for 2 wk failed to produce any treatment-related effects in either sex of exposed rats.

A similar TEA inhalation toxicity study was conducted in mice by the same workers (Mosberg et al. 1985b). Male and female B$_6$C$_3$F$_1$ mice were exposed to 0, 125, 250, 500, 1000, or 2000 mg/m^3 TEA aerosol 6 hr/d, 5 d/ wk for 16 d. The body weights of high exposure group females, but not males, were significantly depressed (21%) relative to controls. A striking, yet variable, response was noted in several hematological parameters in exposed mice. Red blood cell counts, hemoglobin levels, and hematocrit were increased relative to controls; however, these changes lacked dose-response. Rather than representing an unusual treatment-related effect, the authors attributed changes in erythron to a "physiologic response of the animals to hypoxia induced by the relatively high concentrations of the test article during the exposure periods." In contrast, decreased white blood cell counts were also noted in all intermediate, but not high, dose groups of males. Specifically related to changes in lymphocyte numbers, this latter change also lacked dose–response and was not observed in females. These factors, along with the absence of histopathological changes in bone marrow or lymphoid tissues (despite decreased thymus weights), suggested that changes in white blood cell counts reflected normal variability or an unusually low control value rather than a significant treatment-related effect.

Further evidence of the stress of exposure to relatively high levels of aerosolized TEA was observed in mice exposed to ≥ 1000 mg/m^3 TEA aerosol as decreased thymus weights of females relative to controls. Absolute heart weights of females exposed to ≥ 1000 mg/m^3 were also decreased, and heart-to-brain weight ratios were decreased in several intermediate dose groups of females. As with changes in hematological parameters, however, heart weight changes lacked dose–response and, in contrast to high dosage DEA-induced cardiac effects, lacked a histopathological correlate. Indeed, no gross or histopathological changes were observed in any tissues at any exposure level that could account for the observed organ weight changes. As in the rat study, the only histological change observed was a "minimal inflammation of the laryngeal submucosa," the significance of which was questioned because of lack of dose–response and sporadic occurrence within the sections of the larynx examined. Again, no lesions suggestive of severe chemical irritation were reported in the nasal mucosa or pulmonary tissues of exposed mice. Overall, conclusions regarding levels of TEA exposures causing or not causing direct chemically induced toxicity distinct from secondary stress effects and biological variability are difficult to ascertain from this study.

Oral.   The potential oral toxicity of TEA has been examined in several rodent species by "forced bolus dosing" or the addition of TEA to drinking water or feed. In an early study, Kindsvatter (1940) administered 0, 200, 400, 800, or 1600 mg kg^{-1} d^{-1} of commercial or high purity grades of TEA to guinea pigs via forced feeding 5 d/wk and to albino rats via their feed 7 d/wk; sexes were not indicated. One-third of the animals were sacrificed after 12 wk, the second third after 24 wk, and the remaining animals after 24 wk of TEA administration plus a 3-mon recovery period. Results of a number of blood chemistry and urine parameters, gross pathology, and histopathological changes in several tissues were reported. Dose- or time-related changes appeared to be restricted to kidney and liver tissues in both species over the entire dose range. In kidneys, lesions consisted of varying degrees of cloudy swelling of the convoluted tubules and loop of Henle. In livers, lesions consisted of varying degrees of cloudy swelling and occasional fatty changes.

Examination of sciatic nerves of treated animals of both species revealed scattered degeneration in the myelin sheath of the individual fibers; however, these changes lacked dose–response. Rats were generally more sensitive than guinea pigs, and lesions were observed to be reversible following the recovery period. The authors concluded that repeated dosing of rats and guinea pigs with TEA (no specific dosage identified) caused "only slight reversible effects on liver and kidney." In another study, Smyth et al. (1951) administered dosages of 5–2610 mg kg^{-1} d^{-1} TEA to rats (sex not indicated) via their feed for 30 d. Animals ingesting ≥ 80 mg kg^{-1} d^{-1} TEA had reduced growth rates, altered organ weights, microscopic lesions, and increased mortality.

In a recent evaluation of the oral toxicity of TEA, male and female Fischer-344 rats were provided drinking water containing 0%, 0.5%, 1%, 2%, 4%, or 8% TEA for 14 d (Hejtmancik et al. 1985a). Lack of palatability and resultant severe dehydration precipitated the early sacrifice of all but one of the high dose group rats. Dosages for the remaining dose groups, based upon average body weight and water consumption data, were calculated to be approximately 0, 660, 1390, 2570, and 2850 mg kg^{-1} d^{-1} in males and 0, 680, 1380, 2810, and 4380 mg kg^{-1} d^{-1} in females. Some palatability problems were also noted with the 4% TEA drinking water solution. Rats imbibing the latter solution had decreased water consumption, decreased body weight, and organ weight changes secondary to body weight loss. A possible exception may have been the roughly 13% increase observed in the absolute kidney weights of females imbibing 2% TEA solution. Histological evaluation of these and other tissues failed to detect any treatment-related changes. In addition, no treatment-related changes were observed in hematological parameters in rats imbibing TEA solutions of 4% or less. The authors concluded that, under the conditions of the study, no treatment-related changes were observed in male or female rats imbibing drinking water containing 2% TEA or less (approximately 2500–2800 mg kg^{-1} d^{-1} dosages).

In a companion study, male and female B$_6$C$_3$F$_1$ mice were provided drinking water containing 0%, 0.5%, 1%, 2%, 4%, or 8% TEA for 14 d (Hejtmancik et al. 1985b). Actual dosages were not calculated due to potential inaccuracies in the water consumption data. In contrast to rats, no severe palatability problems or mortality were noted in mice imbibing 8% TEA solution. Indeed, the body weights of high dose mice appeared to be relatively unaffected by treatment. Paradoxically, the water consumption of mice imbibing the two highest concentrations of TEA were significantly depressed relative to controls (30%–60%), possibly reflecting the uncertainty about these data. High dose males and females had decreased thymus weights relative to controls, suggesting stress. The only histological change reported, however, was increased hepatocellular "cytoplasmic vacuolization" in the livers of both sexes of high dose group mice. The latter lesion was consistent with "glycogen deposition," possibly related to stress. Finally, no treatment-related changes were reported in hematological parameters in any dose group. It was concluded that no treatment-related changes were observed in either sex of mice imbibing a 4% TEA solution. Assuming roughly normal water consumption, a NOEL of approximately 9000–10,000 mg kg $^{-1}$ d $^{-1}$ TEA was identified.

Dermal.   A relatively complete set of toxicity data has been obtained for dermally administered TEA in rats and mice. Again even though TEA was applied to the shaved backs of animals, an additional route of exposure, possibly the major one, would have occurred orally via ingestion of TEA during grooming. Another factor to consider is the frequent use of neat TEA at all dosages. As noted, this practice increases the potential to cause skin lesions or modification of the skin as a barrier to absorption of TEA, with subsequent systemic effects at artificially low dosages.

The potential repeated dose toxicity of dermally administered TEA has been examined in rats and mice as yet another part of the NTP evaluation of ethanolamines. Male and female Fischer-344 rats were administered dosages of 0, 141, 281, 563, 1125, and 2250 mg kg $^{-1}$ d $^{-1}$ TEA via application of variable amounts of the undiluted test material (neat) to their shaved backs 5 d/wk for a total of 12 doses (Hejtmancik et al. 1985c; Melnick et al. 1988). Both sexes of high dose rats had decreased feed consumption and body-weights, and body-weight-related alterations in the weights of several organs. A necrotizing, chronic-active inflammatory process was noted in the skin at the site of TEA application in both sexes of rats administered ≥ 563 mg kg $^{-1}$ d $^{-1}$; however, no histopathological changes were observed in systemically exposed tissues. Hematological parameters were also not significantly altered in treated animals. In a companion study, male and female $B_6C_3F_1$ mice were administered dosages of 0, 214, 427, 843, 1686, and 3370 mg kg $^{-1}$ d $^{-1}$ TEA neat under a similar study design (Hejtmancik et al. 1985d; Melnick et al. 1988). The only apparent treatment-related effect observed in the latter study was a variable incidence of inflammation in the skin at the site of TEA application in some treated animals. Overall, the NOAEL for the repeated-dose dermal administration of TEA to rats and mice would appear to be 281 and 1686 mg kg^{-1} d^{-1}, respectively, based upon the relatively minimal effects observed in skin at the application site and lack of significant systemic toxicity.

In a recent study, male and female Fischer-344 rats were administered dosages of approximately 0, 125, 250, 500, 1000, or 2000 mg kg^{-1} d^{-1} TEA as an acetone solution or neat chemical via dermal application 5 d/wk for 13 wk (Hejtmancik et al. 1987e; NTP 1994). Significant decreases in body weight gains occurred in high dose group males and females, resulting in terminal body weights roughly 17% and 9% lower than controls, respectively. In addition, the authors reported a "statistically and biologically significant" increase in kidney weights of males and females at dosages of ≥ 500 and ≥ 1000 mg kg^{-1} d^{-1}, respectively. Histological changes were also noted in the kidneys of treated and control females alike but not of males. These changes were characterized by the occurrence of focal or multifocal hyperchromatic basophilic cortical tubular epithelium ("renal tubular regeneration") and tubule mineralization, which were increased in severity or incidence in all dose groups and in females administered ≥ 500 mg kg^{-1} d^{-1} TEA, respectively. Renal tubule mineralization, however, was not dose related, and Hejtmancik et al. (1987e) considered renal effects in general to be "incidental" and not related to treatment.

In contrast, treatment-related gross and histopathological changes, including ulceration, were reported to occur in a dose-related manner in skin at the site of application. Finally, a number of other treatment-related, yet relatively minor, changes in clinical chemistry, hematological, and urinalysis parameters were also observed in higher dose group male and female rats. Overall, the minimal nature of skin lesions at the site of application and lack of dose-related nephropathy in both sexes suggested that the NOAEL for subchronic dermal administration of TEA in rats was 250 mg kg^{-1} d^{-1}.

In a companion subchronic study, male and female B$_6$C$_3$F$_1$ mice were administered TEA dermally 5 d/wk for 13 wk (Hejtmancik et al. 1987f; NTP 1994). Dosages of approximately 0, 250, 500, 1000, 2000, or 4000 mg kg^{-1} d^{-1} were applied to the shaved backs of both sexes as an acetone solution, with the exception of the high dosage, which was applied neat. The body weights of all dose groups of males were decreased relative to controls but not in a dose-related manner, suggesting an unusually heavy control group. No change was noted in female body weights. Despite their lower body weights, the absolute liver and kidney weights of high dose males were elevated roughly 8% and 11% relative to controls, respectively. In females, the liver, kidney, spleen, and heart weights of high dose group animals were elevated about 15-17%, and the kidney weights of all remaining groups of treated females were also variously elevated 6%-8%, irrespective of dosage. Organ weight changes were not, however, accompanied by histopathological changes.

The most consistent systemic change observed was a dose-related decrease in serum sorbitol dehydrogenase activity in both sexes of high dose
mice. The significance of the latter change is questionable given that only
increases in activity are typically associated with potentially untoward
changes in liver membrane function. Skin at the site of application in high
dose group mice of both sexes was observed to be white and crusted. Micro-
scopically, the skin of these latter animals displayed chronic inflammation
and acanthosis. Evidence of acanthosis was noted in all remaining treatment groups. Given the lack of consistent, dose-related, systemic pathological changes and the relatively minor changes in skin at the application site,
the NOAEL for dermally applied TEA in mice would appear to be between
1000 and 2000 mg kg^{-1} d^{-1}.

In another study, approximately 0, 140, 460, or 2000 mg kg^{-1} d^{-1} TEA
and approximately 0, 160, 540, or 2300 mg kg^{-1} d^{-1} TEA in an acetone
vehicle or neat was applied to the shaved backs of male and female C3H/
HeJ mice, respectively, 3 d/wk for 95 d (DePass et al. 1995). Data were
collected on body weight, clinical chemistry parameters, organ weights,
and histopathological changes. No systemic treatment-related effects were
observed in any parameter measured. Skin effects were limited to slight
epidermal hyperplasia at the site of application in mice from all dosage
groups. Consistent with the previous subchronic study (Hejtmancik et al.
1987f; NTP 1994), the NOAEL for dermally administered TEA in this
study appears to be approximately 2000–2300 mg kg^{-1} d^{-1}.

2. Chronic Toxicity

Oral.   The potential chronic toxicity and oncogenicity of orally administered TEA has been evaluated in both rats and mice. Maekawa et al. (1986)
provided male and female rats with drinking water containing 0 (control),
1%, or 2% of a high purity TEA for 2 yr followed by tap water for an
additional 9 wk. The dose levels in females were reduced by half from
week 69 because of excessive treatment-related nephrotoxicity. Based upon
reported body weight and water consumption data, dosages were approximately 525 and 1100 mg kg^{-1} d^{-1} in males and, initially, approximately 910
and 1970 mg kg^{-1} d^{-1} in females. After week 69 of the dosing period,
dosages in females were approximately 455 and 985 mg kg^{-1} d^{-1}. Total
TEA intake over the 2-yr period was calculated to be 170 and 119 g per rat
in males and females, respectively, imbibing the 1% TEA solution, and 358
and 232 g per rat in males and females, respectively, imbibing the 2%
solution.

The most significant treatment-related effects observed in rats chronically imbibing relatively high dosages of TEA were depressed body weights
and histopathological changes in kidneys (Maekawa et al. 1986). Terminal
body weights of treated males and females were decreased as much as 10%
and 14%, respectively, relative to controls. Despite depressed body weights,

absolute kidney weights of male and female rats were increased up to 20% and 58%, respectively, in a dose-dependent manner. Grossly, these tissues were noted to have a granular appearance and were anemic in color. Histopathological changes consisted of an "acceleration of so-called chronic nephropathy" commonly observed in the kidneys of aging Fischer-344 rats. A dose-related increase in the severity and incidence of mineralization of the renal papilla, nodular hyperplasia of the pelvic mucosa, and pyelonephritis with or without papillary necrosis were also observed in the kidneys of both sexes of rats imbibing TEA.

A positive trend was noted in the occurrence of hepatic tumors in males and of uterine endometrial sarcomas and renal cell adenomas in females. Apparent changes in tumor yields, however, were attributed to the unusually low incidence of these tumors in controls relative to laboratory historical values. It was also suggested that the occurrence of renal tumors in high dose female rats may have been a consequence of the extensive kidney toxicity observed in these animals. Indeed, the roughly 14% decreased body weights and increased mortality of high dose females relative to controls (42% vs. 16%) suggest that the maximum tolerated dosage (MTD) was exceeded. The authors concluded that, under the conditions of the study, TEA was not carcinogenic.

The chronic oral toxicity and oncogenicity of TEA has also been evaluated in mice. Male and female $B_6C_3F_1$ mice were provided drinking water containing 0, 1.0, or 2.0% TEA (approximately 98% purity with 1.9% DEA) for approximately 82 wk (Konishi et al. 1992). Total dosages of TEA administered in the study by mice imbibing 1% and 2% solutions were approximately 27–37 and 63–64 g/mouse, respectively. It is estimated that maximum dosages averaged more than 3000 mg kg^{-1} d^{-1} TEA in both sexes of high dose mice. Average body weights of all groups of animals were similar during the dosing period with the exception of high dose males and females, which were depressed during the last 3–4 mon of dosing. Despite this, no significant differences were observed in terminal body weights between treated and control groups of animals. Indeed, no treatment-related differences were noted in organ weights, nor were gross or histopathological changes associated with TEA ingestion noted. A number of tumors were observed in both control and treated animals; however, tumor incidence either lacked statistical significance relative to controls, lacked dose–response, or were well within historical control values. The authors concluded that the study "documented a lack of carcinogenic activity of triethanolamine in $B_6C_3F_1$ mice."

An additional, but highly criticized, oral bioassay of TEA has also been reported (Hoshino and Tanooka 1978). While few experimental details were provided, it appears that male and female ICR/JCL mice were provided diets containing 0, 0.03%, or 0.3% TEA (no purity reported) for their entire life span. Only approximately half of the 40 mice per dose group survived past week 85 of the study. The authors concluded that the total incidence of malignant tumors in females at both dosages was significantly higher than that observed in the controls. In particular, it was claimed that treated females had a higher incidence of lymphoid tissue tumors (19%–25%), primarily thymic lymphoma, than controls (3%). The authors discussed the possibility that NDELA formed in the feed during its preparation (feed was heated for 40 min at 100 °C) or *in situ* in gastric fluid may have been responsible for the observed results. However, the major criticisms of this work have focused on the extremely low incidence of lymphomas and total tumors reported in control mice (a single lymphoma). The tumor yield in this study was in stark contrast to the normally high spontaneous incidence of tumors, particularly lymphomas, which typically can exceed 40% in this inbred strain of mouse (Inai et al. 1979, 1985). For this and other reasons, these data have not been useful or relevant for risk assessment of TEA.

Dermal.   Several bioassays utilizing dermal administration of TEA have been conducted in rats and mice. All appear to suffer the same uncertainties relative to route of exposure as the dermal studies of shorter duration noted earlier. In addition, the potential impact of various methodological, interpretive, and even disease problems upon these studies suggest that caution should be exercised in the use of these data for human risk assessment.

In a recent study, varying amounts of an acetone solution of TEA (55–272 mL/rat) were "painted" on the shaved backs of male and female Fischer-344 rats at dosages of 0, 32, 64, or 125, mg kg^{-1} d^{-1} and 0, 63, 125, or 250 mg kg^{-1} d^{-1}, respectively, 5 d/wk for 2 yr (Hejtmancik et al. 1995; NTP 1994). Based upon mean body weight data, the total estimated lifetime dosages of TEA administered high dose males and females were approximately 24 and 31 g per rat, respectively. Vehicle control groups of rats were administered acetone only; however, no untreated controls were included in the study. It is thus difficult to ascertain the potential effects of the acetone vehicle upon the results of the study, for example, via acetone's effects upon the absorption and skin irritancy of TEA. Satellite groups of rats were sacrificed following 15 mon of dosing. Data were collected on clinical symptoms, body weight, liver and kidney weights (15-mon groups only), and gross and histopathological changes.

The survival rate of high dose females to study termination was less than that of controls, 36% vs. 50%. Not surprisingly, the body weights of these same animals were also depressed by as much as 12% relative to controls during the dosing period and by roughly 7% at terminal sacrifice. In contrast, no significant changes in mortality or body weight were noted in treated males. A number of changes indicative of chronic irritation were observed in skin at the site of application in both sexes of treated rats, including acanthosis, inflammation, ulceration, and erosions. In males, this was restricted to high dose rats, but all dose groups of females were affected.

Systemic effects in rats chronically administered TEA via dermal application were primarily limited to kidney tissues (Hejtmancik et al. 1995; NTP 1994). Kidneys underwent both the "standard" histological evaluation and an additional extensive "step-section" evaluation. The kidney weights of high dose females were elevated roughly 12% relative to controls following 15 mon of dosing. Despite this, no dose-related increase in nontumorigenic histopathological changes was noted in renal tissues of these or any other rats following 15 or 24 mon of dosing. Chronic nephropathy, typically seen in this strain of rat, was observed to a similar degree in nearly all control and treated animals of both sexes at both time points. Microscopic examination of renal tissues also revealed a number of animals with hyperplasia of the tubular epithelium and relatively small microscopic adenomas. The incidence of renal adenomas (8%) observed during the "standard" evaluation of tissues in males administered the intermediate dosage of 63 mg kg^{-1} d^{-1} TEA was numerically higher than in controls and slightly higher than the highest incidence reported in untreated historical controls (0–6%).

Interestingly, evaluation of step-sectioned kidneys from these groups of animals, with its significantly increased power to detect lesions relative to standard sectioning procedures, failed to identify a greater number of tumors in treated rats than the standard method. The total yield of renal adenomas upon combining the results of the two evaluations was 2%, 4%, 12%, and 8% in ascending order of dosages. Combining all proliferative lesions, foci of hyperplasia (a potential precursor to tumor formation) plus adenomas, however, resulted in an almost identical incidence between control and treatment groups of rats (20%–26%). This suggests a lack of a tumorigenic response in the kidneys of male rats. No treatment-related increase in tumor incidence was noted in any other organ system in male rats or in any treatment groups of female rats. Overall, the study failed to generate clear evidence of a carcinogenic response in rats and that the male kidney tumor data were "equivocal."

In a companion study, the potential chronic toxicity and carcinogenicity of dermally administered TEA was evaluated in mice (Hejtmancik et al. 1995; NTP 1994). Varying amounts of an acetone solution of TEA (33–105 mL/mouse) were applied to the shaved backs of male and female $B_6C_3F_1$ mice at dosages of 0, 200, 1000, or 2000 mg kg^{-1} d^{-1} and 0, 100, 300, or 1000 mg kg^{-1} d^{-1}, respectively, 5 d/wk for 2 yr. Based upon mean body weight data, the total estimated lifetime dosages of TEA administered high dose group males and females were approximately 44 and 21 g per animal, respectively. As in rats, an indeterminate amount of the test material would have been ingested during grooming. Vehicle control groups of mice were administered acetone only; however, no untreated controls were included in the study. Thus, as in the rat bioassay, the potential confounding effects of the acetone vehicle are unknown. Satellite groups of mice were sacrificed following 15 mon of dosing. Data were collected on clinical symptoms, body weight, liver and kidney weights (15-mon groups only), and gross and histopathological changes. No significant changes in body weight were noted in treated mice of either sex relative to controls. As in the subchronic study, the skin of mice appeared to tolerate the repeated exposure to concentrated solutions of TEA better than that of rats, as acanthosis and chronic inflammation, but no necrosis, were observed only in high dose mice.

Despite the relatively high dosages of TEA employed in the mouse bioassay, no significant treatment-related changes in mortality or body weight and relatively few systemic effects were observed (Hejtmancik et al. 1995; NTP 1994). The kidney weights of males administered ≥ 630 mg kg^{-1} d^{-1} TEA were elevated as much as 21% relative to controls; however, this change was not associated with histopathological changes in renal tissues. The most significant nonneoplastic changes were found in the livers of male mice consistent with a chronic bacterial hepatitis in these animals (hepatocyte karyomegaly and oval cell hyperplasia). Selective staining of hepatic tissues revealed the presence of a *Helicobacter* species, presumably *H. hepaticus*, which has been shown by Ward et al. (1994a) to cause acute focal necrosis and inflammation in mice followed by chronic regenerative DNA synthesis and the symptoms noted in the present study. Infection has been shown to be associated with a higher incidence of hepatocellular neoplasms in mice (Ward et al. 1994b). No overt histological evidence of bacterial infection was observed in hepatic tissues of most female mice following 15 and 24 mon of dosing; however, subsequent analysis revealed hepatitis in these mice as well (see following).

The issue of potential tumorigenicity of chronically administered TEA in the mouse bioassays has primarily focused on changes in the incidence of benign tumors of the liver and their potential relationship with bacterial hepatitis or obesity in treated animals. The incidence of hepatic adenomas in male and female mice administered 2000 and 1000 mg kg^{-1} d^{-1} of TEA, respectively, were observed to be numerically higher than in controls (54% vs. 74% and 44% vs. 80%, respectively). The incidence of carcinomas was also elevated, but only in females and not in a dose-related manner. Significantly, the incidence of liver tumors in high dose male mice was closely linked with histological evidence (see earlier discussion) of bacterial infection. In fact, 94% of the mice with evidence of infection also had tumors. In recognition of this "confounding factor," the authors concluded that these data did not present "clear evidence" of a carcinogenic response and considered the response in male mouse livers to be "equivocal."

As in male mice, a correlation between liver tumors and active bacterial hepatitis in high dose female mice as part of the NTP study has been clearly identified in a subsequent conduct of the study. Direct presence of the bacteria in liver tissues of the few mice examined and typical histopathology indicative of an active chronic hepatitis were not observed in the study (Hejtmancik et al. 1995; NTP 1994). However, direct culturing, fluorescence antibody binding, and PCR amplification product analyses revealed *H. hepaticus* in frozen liver tissue (J.G. Fox, personal communication). It is of interest that the unusually high incidence of adenomas observed in the control females (44%) was more than double that observed in the only other dermal study with an acetone vehicle conducted at the testing facility (16%). It has been suggested that liver tumor rates in mice display a strong association with body weight, especially at 12 mon of age (Seilkop 1995). It is thus noteworthy that 12-mon-old females in the present study had higher body weight than comparable females in the historical database study (average, 46 vs. 36 g). Based upon the historical data presented by Seilkop (1995), however, a somewhat lower tumor incidence, roughly 33%, would have been expected in this study. The original conclusion that the liver tumor data in female mice, while not clear evidence of a tumorigenic response, still represented "some evidence" of carcinogenic activity of TEA was made prior to the finding of the *H. hepaticus* infection in these animals.

An additional chronic dermal bioassay of TEA has been conducted in which the potential tumorigenicity of dermally administered "pure" TEA (99% purity) and "technical" TEA (80% purity) has been examined in mice (Kostrodymova et al. 1976). Few details of this Soviet study are available, but it appears that TEA was applied to the backs of male CBA × C57Bl6 mice as a 50% solution in acetone twice weekly for 14–18 mon (dosages or dosing volume not given). As part of the overall bioassay, the potential of TEA to act as a promoter of 3-methylcholanthrene- (3-MC-) initiated skin tumors was also evaluated. In the apparent initiation-promotion portion of the study, appropriate groups of mice were also dermally administered 3-MC (dosage not given) prior to TEA dosing and following 3 wk of the dosing period. Examination of the dermal site of application and of major organ systems indicated no gross nontumorigenic changes related to dosing. The few systemic tumors reported were distributed roughly evenly between treated and control groups of mice. The incidence of skin tumors in mice treated with 3-MC followed by TEA was statistically higher than in mice treated with 3-MC alone (approximately 2.5- to 4-fold following 35–75 wk of dosing). These data suggest a possible promoter activity in skin for dermally applied TEA which, at least in some dermal studies, have induced irritant-related hyperplasia of epidermal tissues. Overall, however, the authors concluded that there was no direct tumorigenic activity of TEA. Finally, TEA was negative in the TG.AC transgenic mouse bioassay at a dermal dose of 30 mg/mouse/d, 5 d/wk (Tennant et al., 1995).

VII. Genotoxicity

Mono-, di-, and tri-alkanolamines have been evaluated for genotoxic activity in a broad spectrum of *in vitro* and *in vivo* assays. These have included assays of the potential of these compounds to cause mutations in bacteria and *Drosophila*, damage DNA in bacteria and cultured mammalian cells, induce the transformation of mammalian cells in culture, and cause chromosomal damage in a variety of mammalian cells, yeast, and even plants. As shown in Table 15, the results of these tests have been nearly uniformly

Table 15. Genotoxicity data for mono-, di-, and tri-ethanolamines.

Compound	Test	Results	Reference
Monoethanolamine			
Mutagenicity	*Salmonella* (various Ames tester strains with or without metabolic activation)	Negative	Dean et al. (1985); Hedenstedt (1978); Mortelmans et al. (1986)
	Salmonella ("Ames systems", no details available)	"Weak mutagenic effects"	Arutyunyan et al. (1987)
	Escherichia coli WP2 tyr⁻ (with or without metabolic activation)	Negative	Dean et al. (1985)
	Saccharomyces cerevisiae (with or without metabolic activation)	Negative	Dean et al. (1985)
Cell transformation	Hamster embryo (with feeder cells)	Negative	Inoue et al. (1982)
Clastogenic assayss	Rat liver cell line RL$_4$ chromosomal aberration	Negative	Dean et al. (1985)
	Human lymphocytes (no details available)	"Weak" positive	Arutyunyan et al. (1987)
	Plant (*Crepis capillaris* seeds) chromosomal aberration (no details available)	"Weak" positive	Arutyunyan et al. (1987)
Diethanolamine			
Mutagenicity	*Salmonella* (various Ames tester strains with or without metabolic activation)	Negative	Dean et al. (1985); Haworth et al. (1983); Hedenstedt (1978); Melnick (1992)
	E. coli WP2 tyr⁻ (with or without metabolic activation)	Negative	Dean et al. (1985)
	Mouse lymphoma L5178Y/TK +/−	Negative	Melnick (1992); Myhr et al. (1986)

Macromolecular reactivity assay		
4-(p-Nitrobenzyl) pyridine alkylation	Negative	Hedenstedt (1978)
Cell transformation		
Chinese hamster embryo (with feeder cells)	Negative	Inoue et al. (1982)
Clastogenic assays		
Chinese hamster lung chromosomal aberration	Negative	Dean et al. (1985)
Mouse micronucleus	Negative	Melnick (1992)
Rat liver cell line RL_4 chromosomal aberration	Negative	Dean et al. (1985)
Chinese hamster ovary sister chromatid exchange (with or without metabolic activation)	Negative	Loveday et al. (1989); Melnick (1992)
Chinese hamster ovary chromosomal aberration (with or without metabolic activation)	Negative	Loveday et al. (1989); Melnick (1992); Sorsa (1988)
Triethanolamine		
Mutagenicity		
Salmonella (various Ames tester strains with or without metabolic activation)	Negative	Inoue et al. (1982); NRC (1981); Dean et al. (1985); Mortelmans et al. (1986); NTP (1994); Pathak et al. (1982); Wang et al. (1988)
Salmonella ("Ames systems", no details available)	"Weak mutagenic effects"	Arutyunyan et al. (1987)
Escherichia coli WP2 tyr⁻ (with or without metabolic activation)	Negative	Inoue et al. (1982); Dean et al. (1985)
Drosophila melanogaster sex-linked recessive lethal	Negative	Yoon et al. (1985); NTP (1994)

(continued)

Table 15. (*Continued*)

Compound	Test	Results	Reference
DNA damage repair assays	*Bacillus subtilis* Rec (with or without metabolic activation)	Negative	Inoue et al. (1982); Hoshino and Tanooka (1978)
	Primary rat hepatocyte UDS	Negative	Litton Bionetics (1982); Beyer et al. (1983)
Cell transformation	Hamster embryo (with feeder cells)	Negative	Inoue et al. (1982)
Clastogenic assays	Chinese hamster lung chromosomal aberration	Negative	Inoue et al. (1982)
	Rat liver cell line RL$_4$ chromosomal aberration	Negative	Dean et al. (1985)
	Chinese hamster ovary sister chromatid exchange (with or without metabolic activation)	Negative	Galloway et al. (1987); NTP (1994)
	Chinese hamster ovary chromosomal aberration (with or without metabolic activation)	Negative	Galloway et al. (1987); NTP 1994
Triethanolamine			
Clastogenic assays	Human lymphocytes (no details available)	"Weak" positive	Arutyunyan et al. (1987)
	Drosophila melanogaster reciprocal trans-location	Negative	Yoon et al. (1985)
	Sacchraromyces cerevisiae (with or without metabolic activation)	Negative	Dean et al. (1985)
	Plant (*Allilum cepa*) multipolar mitosis	Negative	Barthelmess and Elkabarity (1962)
	Plant (*Crepis capillaris* seeds) chromosomal abberration (no details available)	"Weak" positive	Arutyunyan et al. (1987)
	Plant (*Cyanopsis tetragonoloba*) chromosomal aberration (no quantitative data available)	Positive "meiotic irregularities"	Bose and Naskar (1975)

negative, indicating a lack of genotoxicity. Reports of positive findings, primarily chromosomal alterations in several plant assays, have lacked experimental detail or quantitative data necessary for critical evaluation.

A. Monoethanolamine

Ethanolamine lacks mutagenic potential in Ames bacterial mutagenicity assays either when plated directly or when preincubated prior to plating with a variety of *Salmonella typhimurium* tester strains developed to identify base-pair substitution or frameshift mutagens (Dean et al. 1985; Hedenstedt 1978; Mortelmans et al. 1986). MEA also failed to cause mutations in a test organism that is sensitive to oxidative-type mutagens, *Escherichia coli* WP2 (Dean et al. 1985). The addition of a postmitochondrial liver fraction recovered from polychlorinated biphenyl- (PCB-) induced rats or hamsters to either assay had no effect. Experimental and quantitative data were not available to evaluate a Soviet report of "weak mutagenic effects" in an "Ames system" (Arutyunyan et al. 1987).

Several assays of the potential of MEA to damage DNA in a bacterial tester strain (*Bacillus subtilis* Rec assay) and to cause chromosomal damage in yeast cells (*Saccharomyces cerevisiae* gene conversion assay) have been negative. MEA was also found to lack any clastogenic activity in cultured rat liver epithelial-like cells (Dean et al. 1985); however, "weak" positive results were reported in cultured human lymphocyte and plant chromosomal aberration assays (Arutyunyan et al. 1987). Again, experimental and quantitative data were not available to evaluate the latter two Soviet studies.

B. Diethanolamine

As in the case of MEA, DEA has been thoroughly evaluated for mutagenic potential in a number of bacterial mutation assay systems (Dean et al. 1985; Haworth et al. 1983; Hedenstedt 1978). In addition, DEA has been shown (Myhr et al. 1986) to be negative in the mouse lymphoma L5178Y mammalian cell mutagenicity assay. DEA also failed to cause the transformation of another mammalian cell type, Chinese hamster embryo cells, to a more anaplastic state *in vitro* (Inoue et al. 1982). The addition of liver enzymes isolated from rats or hamsters treated with PCBs to these assays did not alter the negative responses obtained.

The potential of DEA to cause chromosomal damage has been extensively evaluated. Clastogenesis assays of DEA have been carried out *in vitro* in a number of test organisms ranging from yeast to cultured cells derived from ovary, lung, and liver tissues (Dean et al. 1985; Loveday et al. 1989; Melnick 1992). Results of these tests have been uniformly negative with or without the addition of metabolic fractions recovered from PCB-induced rat liver. DEA has also failed to demonstrate clastogenic activity *in vivo* in a mouse micronucleus test conducted using animals that had been administered up to 1250 mg kg^{-1} d^{-1} DEA via skin painting for 13 wk (Melnick 1992).

While purified DEA has been shown to lack genotoxic potential, it is important to note that, like many secondary amines, it may react chemically with nitrosating compounds under favorable conditions (e.g., low pH and heat) to form a nitrosamine, in this case NDELA. NDELA and a number of its metabolites have been shown to be mutagenic in a variety of short-term genotoxicity assays (see review in ECETOC 1990) and, as noted later, are tumorigenic when administered to test animals.

C. Triethanolamine

As with the other ethanolamines, TEA has been extensively evaluated for mutagenicity in a variety of bacterial assays, including most of the Ames tester strains and *E. coli* WP2 with and without the addition of PCB-induced rat or hamster liver enzymes (see Table 15). Nearly all were negative; however, a weakly positive response was reported in a Soviet "Ames system" for which few details are available (Arutyunyan et al. 1987). TEA also failed to induce sex-linked recessive lethal mutations or reciprocal translocations in several *Drosophila* assays (NTP 1994; Yoon et al. 1985).

As shown in Table 15, TEA failed to damage DNA in the *B. subtilis* Rec and primary rat hepatocyte unscheduled DNA synthesis assays, did not cause the anaplastic transformation of mammalian cells in a hamster embryo cell transformation assay, and did not cause chromosomal changes in a number of test systems ranging from yeast cells to a variety of mammalian cells (Dean et al. 1985; Galloway et al. 1987; Inoue et al. 1982). A weak potential to cause chromosomal damage in cultured human cells and in a plant assay was reported in the same Soviet study noted previously for which few details are available (Arutyunyan et al. 1987). In another abbreviated report lacking quantitative data, it was claimed that adult cluster bean plants sprouted from seeds soaked for 8 hr in a 1% or 2% solution of TEA had an increased incidence of "meiotic irregularities" (Bose and Naskar 1975). In contrast, however, Barthelmess and Elkabarity (1962) reported no increased incidence of multipolar mitoses in *Allium cepa* cells after 4 hr of exposure to an approximately 2% solution of TEA.

VIII. Reproductive and Developmental Toxicity
A. General

The potential of the alkanolamines for developmental toxicity was evaluated in a screening assay originally developed by Chernoff and Kavlock (1982). The assay consisted of three experimental phases. The first two phases were dose range-finding studies to identify the LD$_{10}$, to be used in phase 3. All the treatments were administered by gavage. This method screens chemicals for embryonic, fetal, and neonatal toxic responses using pregnant mice treated during the major period of organogenesis (days 6–15 of gestation). Evaluation of the developmental toxicity potential included maternal body weight, maternal mortality and signs of toxicity, pup counts at birth, pup weight, and offspring survival from birth to day 3 postpartum.

Results (EHRT 1987) indicated that MEA at 850 mg kg^{-1} d^{-1} was toxic to the pregnant animal, producing 16% mortality. The number of viable litters was significantly reduced, but the litter size, the percentage survival of the pups, the birth weight of the pups, and the weight gained by the pups were unaffected. Because of the reduced number of viable litters in conjunction with moderate maternal mortality, MEA was judged to be positive in this preliminary screen.

DEA at 450 mg kg^{-1} d^{-1} had no effect on maternal mortality, litter size, or birth weight of the pups but did decrease the number of viable litters, the percent survival of the pups, and the pup weight gain. DEA was judged to be positive in the preliminary developmental toxicity screen. TEA at 1125 mg kg^{-1} d^{-1} had no effect on any of the parameters monitored. Since the dose tested did not result in maternal mortality, the developmental toxicity potential of TEA could not be appropriately judged.

In addition, the results from this screen were used in conjunction with a scoring system to numerically rank the chemicals for conventional developmental toxicity testing (York et al. 1988). Five indices of potential developmental toxicity were assigned point values that varied according to maternal mortality. These indices were the proportion of pregnant survivors that produced a litter of at least one live-born pup, average litter size and pup weight at birth, and average pup survival and weight gain to 3 d. This scoring method produced a low priority classification for further testing for MEA, a high priority for DEA, and an intermediate priority for TEA.

B. Monoethanolamine

The embryopathic effects of high doses of MEA were evaluated in pregnant Long–Evans rats (Mankes 1986). MEA was given as aqueous solutions by gavage at levels of 0, 50, 300, or 500 mg/kg from day 6 to day 15 of gestation. The dosing volume was held constant at 20 mL/kg. All animals were observed daily for mortality and signs of intoxication. All dams were sacrificed on day 20 of gestation and their fetuses delivered by Cesarian section. Total uterine weight, total litter weight, individual pup weights, crown–rump length, number of live pups, stillbirths and resorptions, implantation sites, sex distribution, and number of corpora lutea were recorded for each pregnancy. All live pups were examined for gross malformations at birth. In addition, the live offspring were classified according to contiguity with offspring of the same or opposite sex, and analyses were conducted with reference to the treatment group and fetal subtype for body weight and the incidence of anomalies.

Signs of mild maternal toxicity (initial lethargy followed by hyperactivity and agitation) were observed in the group of rats treated with 500 mg/kg. All animals appeared to be normal 8 hr after dosing. There were no adverse effects in maternal food consumption or body weight. Significant dose-related increases in embryonic death and fetal malformation and a decrease in mean fetal weight were reported. Thirty percent of the 500 mg/kg fetuses appeared to be runted (defined as a fetal weight less than 2.7 g on the 20th day of gestation) and hydronephrotic, and many had sternebral and rib defects. Forty-six percent of the 300 mg/kg fetuses were grossly abnormal; runting and hydronephrosis/hydroureter were the most common anomalies observed. Skeletal defects of these fetuses included rib, sternebral, and vertebral variations. Thirty-two percent of the 50 mg/kg fetuses appeared grossly abnormal; variant sternebral ossifications were the most common defects observed. However, the number of malformed fetuses per dam was significantly increased only in the 300 mg/kg group, perhaps owing to significant embryolethality in the 500 mg/kg group. Malformed fetuses were also considered as a percent of the litter affected. There appeared to be a significant increase in malformed offspring from all the MEA-treated dams as compared to controls, with pups contiguous to a male pup more adversely affected.

While the results of the Mankes (1986) study suggested that a potential effect might occur at maternally toxic doses, they did not indicate a definitive or significant developmental effect, for a number of reasons. First, the Mankes (1986) study was not a conventional developmental toxicity study but was designed to examine the role of *in utero* position in modulating developmental effects. Analyses of the effect data were modified to more specifically evaluate the effects on fetal subtypes rather than using the litter as the unit of analysis. The latter is typically done in traditional developmental toxicity studies because fetuses within a litter are not considered to be independent of one another. Also, the classification of certain fetal effects as malformations was atypical (e.g., runting and hydroureter). Finally, no dose–response relationship was observed for any of the malformations reported.

To better assess the developmental toxicity potential of MEA in the conventional definition, a developmental toxicity study employing a standard testing protocol was conducted (Hellwig and Hildebrand 1994). In this study, pregnant Wistar rats were treated by gavage with aqueous solutions of MEA at levels of 0, 40, 120, or 450 mg/kg. The dose volume was 10 mL/kg, and the duration of exposure was from day 6 through 15 of gestation. On day 20 postcoitum, the first 25 rats/group were sacrificed and potential toxic effects assessed by gross pathology. The fetuses were examined for any external, soft tissue, and skeletal abnormalities. The remaining animals (15/group) were allowed to litter and rear their pups up to day 21 postpartum, when they were sacrificed and examined macroscopically for any lesions.

There were no treatment-related effects on the dams of the low- and mid-dose groups. Significant reductions in food consumption, mean body weight, and body weight gain at various times of the exposure were observed in the dams in the 450 mg/kg group. However, no effects on any gestational parameters and no signs of developmental toxicity were observed. In contrast to the Mankes (1986) study, neither the fetuses nor the pups showed any increased malformation rate or growth retardation. Thus, MEA was not embryotoxic or teratogenic in the rat following gavage exposure up to and including 450 mg kg^{-1} d^{-1}.

In another developmental toxicity study, Sprague–Dawley rats were exposed dermally to MEA for 6 hr/d at 0, 10, 25, 75, or 225 mg kg^{-1} d^{-1} on days 6 through 15 of gestation, and New Zealand white rabbits were exposed at 0, 10, 25, and 75 mg kg^{-1} d^{-1} on days 6 through 18 of gestation (Liberacki et al. 1996). At Cesarian section on day 21 (rats) or day 29 (rabbits), weights of the maternal liver, kidneys, and gravid uteri were recorded as well as the number of corpora lutea, the number and position of implantations and resorptions, and the number of live or dead fetuses. Fetuses were weighed and examined for external, visceral, and skeletal alterations. Dermal exposure of pregnant rats to 225 mg kg^{-1} d^{-1} and rabbits to 75 mg kg^{-1} d^{-1} resulted in severe skin irritation or lesions. The dermal irritation began with erythema and progressed to necrosis, scabs, and scar formation. In addition, rats given 225 mg kg^{-1} d^{-1} exhibited a significant decrease in body weight gain over the exposure period. Despite the maternal effects observed, there was no evidence of developmental or fetal toxicity at any dose level. Thus, MEA was not embryotoxic or teratogenic following dermal application at exposure levels up to 225 mg kg^{-1} d^{-1} in rats and 75 mg kg^{-1} d^{-1} in rabbits.

C. Diethanolamine

In a range-finding developmental toxicity study (EHRT 1990), Sprague–Dawley rats were given aqueous solutions of DEA by gavage at levels of 0, 50, 200, 500, 800, or 1200 mg/kg from day 6 to day 15 of gestation. The dosing volume was held constant at 5 mL/kg. Fetuses were delivered by Cesarian section on day 20 of gestation. The number of implantation sites, resorptions, dead or live fetuses, and the gravid uterine weight were recorded. All animals at the 500 mg/kg or higher level died or were moribund and sacrificed. No maternal mortality was observed in the 50 or 200 mg/kg groups. Maternal body weight gain was significantly reduced in the 200 mg/kg group. At scheduled sacrifice, a litter was found to be completely resorbed in one dam in the 200 mg/kg group. However, none of the recorded gestational parameters were significantly different between the treatment groups and controls.

Neeper-Bradley (1992) and Neeper-Bradley and Kubena (1993) evaluated the developmental toxicity of DEA to CD rats and New Zealand white rabbits. Aqueous solutions of DEA at 0, 150, 500, and 1500 mg kg^{-1} d^{-1} were administered cutaneously from days 6 to 15 of gestation to rats and at dose levels of 0, 35, 100, or 350 mg kg^{-1} d^{-1} from days 6 to 18 of gestation to rabbits. Prior to scheduled necropsy on day 21 (rats) or day 29 (rabbits) of gestation, blood was obtained for hematological evaluation. Fetuses were evaluated for any external, visceral, and skeletal anomalies.

In the rabbit, DEA at 350 mg kg^{-1} d^{-1} produced marked skin irritation. There were no treatment-related effects on gestational body weights, body weight gains, or hematological parameters in any group. The absolute and relative liver weight and relative kidney weight were increased (7%–16%) in the 350 mg kg^{-1} d^{-1} group, although the increases were not statistically significant. There was no evidence of developmental toxicity in rabbit fetuses at any dose level, and there were no apparent effects of treatment on the incidences of external, visceral, or skeletal abnormalities.

In the rat, DEA at 500 and 1500 mg kg^{-1} d^{-1} produced moderate and severe skin irritation, respectively. Maternal body weight gain was decreased in the 1500 mg kg^{-1} d^{-1} group. Absolute and relative kidney weights were increased at 500 and 1500 mg kg^{-1} d^{-1}. Hematological effects including anemia, abnormal red cell morphology (poikilocytosis, anisocytosis, polychromasia), and decreased platelet count were observed in all treatment groups. The 1500 mg kg^{-1} d^{-1} group also had increased lymphocytes and total leukocytes. In the fetuses, there were no effects of treatment on body weight or on incidence of external, visceral, or skeletal abnormalities. However, increased incidences of six skeletal variations involving the axial skeleton and distal appendages were observed in litters from the 1500 mg kg^{-1} d^{-1} group. The skeletal variations included poor ossification in the parietal bones, cervical centrum #5, and thoracic centrum #10, lack of ossification in all proximal hindlimb phalanges and some forelimb metacarpals, and callused ribs.

No embryotoxic or teratogenic effects were produced by topical administration of 2 mL/kg semipermanent hair dye preparations containing 2% DEA (equivalent to about 40 mg/kg DEA) to the shaved backs of pregnant Charles River CD rats on gestation days 1, 4, 7, 10, 13, 16, and 19 (Burnett et al. 1976).

D. Triethanolamine

The embryotoxic effects of TEA were tested in the chicken embryo assay (Korhonen et al. 1983). Three-day-old white Leghorn chicken eggs were injected with 0.5–4.0 μmol of TEA in acetone. Eleven days after the injection, eggs were opened and the embryos inspected for survival and external malformations. The ED$_{50}$ for embryotoxic effects was 2.6 μmol/egg. The embryotoxic effects included early mortality and malformations (open coelom, short back or neck, edema, and lymph blebs). The incidence of malformations (3%–6%) in the TEA-treated groups was not significantly different from that of controls.

In mating trial studies using male and female Fischer-344 rats, 500 mg/

kg of TEA in acetone was applied dermally to the interscapular area of the clipped back in a volume of 1.5 mL kg^{-1} d^{-1} for 10 wk prior to mating, during breeding, and through gestation and lactation for females. No effects on mating, fertility, or offspring growth and survival were observed (Battelle 1988a). In similar studies with Swiss CD-1 mice administered daily applications of 2000 mg/kg in a volume of 3.6 mL/kg, no chemical-related effects occurred other than ruffled fur in females and irritation at the application site of males and females (Battelle, 1988b).

No embryotoxic or teratogenic effects were produced by topical administration of 2 mL/kg semipermanent hair dye preparations containing 0.1%–1.5% TEA (equivalent to about 2–30 mg/kg TEA) to the shaved backs of pregnant Charles River CD rats on gestation days 1, 4, 7, 10, 13, 16, and 19 (Burnett et al. 1976).

The summary conclusions on the developmental and reproductive toxicology of the alkanolamines is presented in Table 16. The available scientific evidence suggests that neither MEA, DEA, nor TEA is teratogenic. There is also no evidence of embryofetal toxic effects associated with exposures to MEA or TEA. With DEA, there are some indications of embryofetal toxicity manifested as increased incidences of some skeletal variations. However, the rat fetuses were much less susceptible than their mothers to the systemic effects of topically applied DEA. Thus, the observed develop-

Table 16. Developmental and reproductive toxicology summary of alkanolamines.

	MEA	DEA	TEA
	No-observable-effect level (mg kg^{-1} d^{-1})		
Reproductive toxicity screen			
Mouse, dermal	–	–	>2000
Rat, dermal	–	–	>500
Developmental toxicity studies			
Maternal toxicity			
Rat, gavage	120	50	–
Rat, dermal	75	<150	>30
Rabbit, dermal	25	100	–
Embryofetal toxicity			
Rat, gavage	>450	>200	–
Rat, dermal	>225	500	>30
Rabbit, dermal	>75	>350	–
Teratogenicity			
Rat, gavage	>450	–	–
Rat, dermal	>225	>1500	>30
Rabbit, dermal	>75	>350	–

mental effects with DEA would not be expected to occur at exposure levels that are not toxic to the mother.

IX. Discussion

The dual functional groups, amino and hydroxyl, of the ethanolamines (aminoethanols) make them useful in a large number of industrial applications (i.e., solvents, intermediates in the production of soaps, surfactants, and salts, gas and textile processing, corrosion inhibitors, cement and concrete additives, and pharmaceuticals). In most of these applications, the basic amino group reacts with organic or inorganic acids to form neutral amides or water-soluble salts. Fatty acids, such as lauric acid, react with the alcohol functional group of DEA to form esters. Alkanolamines are used per se in cutting fluids and in cosmetics. Under conditions of use, exposure (dermal or inhalation) is believed to be at a low level and in most cases of short duration.

The bifunctional nature of MEA plays a major role in the biological fate and toxicity of this compound in man and animals, where the alcohol group is phosphorylated and phosphorylated ethanolamine is transferred to cytidine monophosphate to form CDP-alkanolamine and phospholipids via diacylglycerol. The importance of MEA (ethanolamine) as a structural component of phospholipids (headgroups in phospholipid bilayers) and the role of these lipids in the formation of biomembranes (composed of lipid and protein) that form the outer surface of all cells and divide cells into internal compartments cannot be overemphasized (Sanger and Nicholson 1972).

In biomembranes, proteins recognize the ester groups and headgroups such as choline, ethanolamine, and serine of the phospholipids and the hydroxyl and amide functions of the sphinomyelins. In addition to protein-lipid interactions, water molecules are hydrogen bonded to two different anionic oxygen atoms of phosphate groups in adjacent unit cells, forming an infinite phosphate–water hydrogen-bonded ribbon. The three-dimensional structure of these biomembranes is currently being examined (Kuo 1985; White 1994). Membrane phospholipids play an active role in signal transduction and functional modulation. Proteins embedded in outer cell membranes (ATPase or ABC transporters) regulate the movement of chloride ions and nutrients across cell membranes. This group of proteins also includes drug-resistant protein that ejects chemotherapeutic drugs from cancer cells (Kartner and Ling 1989).

Adipose tissue makes up approximately 6% of the body of the rat, with 85.5% of the weight being neutral lipid and 2.5% phospholipid. The body of a rat weighing 250 g contains 15 g of adipose tissue, 12.83 g of neutral lipid, and 0.375 g (375 mg) of phospholipid (PE, PC, PS, etc.). Choline is obtained from dietary sources or formed *de novo* by the methylation of phosphatidylethanolamine. The liver of a typical rat contains 6% lipid by weight, 2.5% phospholipid, and 3.5% neutral lipid. According to Sundler (1973), the phospholipid pool in the liver of a rat is approximately 55 mg. Smaller amounts of ethanolamine (61 μg), phosphorylethanolamine (540 μg), and CDP-ethanolamine (106 μg) are present in the liver.

The rate of synthesis of phosphatidylethanolamines from CDP-ethanolamine was reported by Sundler (1973) to be of the order of 0.06–0.08 μmoles/min per liver ($\sim$3.0 mg/hr per liver). Considering this rate of synthesis, 18.0 hr is required to replace all the phospholipid in the liver and 125 hr to replace total body phospholipid. A very small amount of ethanolamine (37.5 mg) is required to replace total body phospholipids. Ethanolamine/choline comprises approximately one-tenth of the molecular weight of the average phospholipid. According to Horrocks (1969), the half-lives of brain ethanolamine/choline phosphoglycerides are roughly 3 d. On this basis, the estimated rat daily dietary requirements for choline would be of the order of 10 mg/kg. Human diets supply 713 mg of choline per 70 kg bw/d (Zeisel 1989). Amounts exceeding 10 mg/kg are metabolized to natural products, CO_2, ammonia, and acetaldehyde. MEA is rapidly incorporated into phospholipids according to Tinoco et al. (1970), who found 27.1% of an administered dose of MEA in phospholipids 10 min after dosing. The studies of Smyth et al. (1951) provided rats with diets of MEA containing 160–2670 mg kg^{-1} d^{-1} for 30 d. These amounts greatly exceed the nutritional requirements of the rat for ethanolamine/choline.

The administration of large amounts of MEA may create an imbalance in the amount of available phosphatidylcholine impairing renal function. According to Michael et al. (1975), choline-deficient animals have abnormal concentrating ability, free water reabsorption, glomerular filtration rate, and renal plasma flow, and gross renal hemorrhaging. MEA per se is highly basic (pH$_a$ of 9.5) and extremely irritating to skin. Changes on the skin and in the lungs of animals topically administered or inhaling vaporized or aerosolized MEA (per se) appear to be related to the extreme irritancy of MEA and the necessity of body tissues to neutralize this strong base to escape toxic effects. MEA appears to have a direct effect on cell membranes in the skin and lungs as well as on membranes in the nasal passages, liver, and kidney. Histopathological changes in a skin include vacuolation of epithelial cells, thickening of the epithelium, inflammation, and necrosis (Weeks et al. 1960). MEA is not readily absorbed through rat, rabbit, or human skin ($>$2% of applied dose in 6 hr; 0.009–0.025 mg cm^{-2} hr^{-1}) (Sun et al. 1996).

Panolobular "fatty metamorphosis" noted in the livers of rats and guinea pigs exposed to MEA vapor (Weeks et al. 1960) may be related to the amount of choline available for the formation of very low density lipids (VLDL) for exporting lipid out of the liver. High concentrations of MEA at the surface of phospholipid bilayers may produce gaps or holes in membranes. According to Hui et al. (1995), there are more defects in monolayers at pH 5 and 9 than at pH 11. Studies on the effects of MEA in water on membrane structure and stability are needed. MEA was not genotoxic in published studies nor was it embryotoxic or teratogenic in the rat following gavage exposure up to and including 450 mg/kg from days 6 through 15 of gestation.

DEA does not occur naturally in animal phospholipids. Oral administration of DEA resulted in its deposition in liver (27%) and kidney (5%), with lesser amounts (<1.0%) being found in blood, brain, spleen, and heart. Long-term dietary administration of DEA decreases the formation of choline-containing phospholipids (Artom et al. 1949). At high substrate concentrations (K_i, 2600–2900 μM), DEA competitively inhibits the incorporation of MEA into phospholipids (Barbee and Hartung 1979a). The K_m for the incorporation of DEA into phospholipids is 11,600 μM, while the K_m for MEA is 53.5 μM. A single oral dose of 250 mg/kg of DEA did not inhibit MEA incorporation. Multiple oral doses (320 mg kg^{-1} d^{-1}) in drinking water over a 3-wk period were required to inhibit the incorporation of ethanolamine and choline into phospholipids (Barbee and Hartung 1979a). *In vivo* studies by Barbee and Hartung (1979b) also indicated that DEA had to be administered at high concentrations for an extended period of time to produce an effect on mitochondrial function and structure.

Numerous changes in subcellular organelles (i.e., mitochondria, endoplasmic reticulum, etc.) were observed in the liver of the mouse by electron microscopy after administration of an oral LD_{50} dose of DEA (2.3 g/kg) by Blum et al. (1972). Analysis of tissues and urine indicated that DEA is not readily metabolized to other materials and is strongly retained by the liver. The high concentration of DEA in liver was first believed by Barbee and Hartung (1979b) to be responsible for changes in mitochondrial membrane ionic permeability. Barbee and Hartung (1979b), however, were unable to demonstrate an increase in membrane permeability over controls in their studies involving NADH and cytochrome c and presumed the effects were related to the involvement of DEA with phospholipid metabolism. Additional studies are necessary to determine the effects of DEA headgroups (PC fraction) and *N*-methyl and *N,N*-dimethyl DEA headgroups (PC fraction) on the structure and stability of phospholipid bilayers. The DEA and methylated DEA headgroups of the phospholipid bilayer may change the manner in which nitrogen atoms hydrogen bond to adjacent phosphate groups or the ability of proteins to recognize phospholipid structure (i.e., PE, PC, or PS). In the case of phospholipids involving PC, the only groups available for hydrogen bonding are the anionic phosphate oxygens and the carbonyl oxygen atoms of the acyl ester groups.

In PE- and PS-containing phospholipids, the nitrogen atom of the ethanolamine chain can be directly involved in N–H . . . O hydrogen bonding with adjacent phosphate groups. Hui et al. (1995) used atomic force microscopy (AFM) to study the effects of headgroups (PE and PC), acyl chain order (saturated and unsaturated), and pH 5, 9, and 11 on the stability of phospholipid monolayers. AFM provides a way to measure monolayer stability and to observe defects resulting from instability. High concentrations of free DEA in the cytosol may affect membrane stability by substituting for MEA and choline in headgroups or by changing the pH in the aqueous layer surrounding PE/PC headgroups. DEA has a pK_a of 8.8; aqueous solutions of DEA approach a pH of 11. In any case, additional work is needed to determine the effects of replacing MEA with DEA and choline with methylated DEA in the headgroups of membrane bilayers.

Degenerative changes in renal tubular epithelial cells, demyelination of nerve tracts in medullary brain and spinal cord tissues, cytoplasmic vacuolization, and degenerative changes in centrilobular hepatocytes and in seminiferous epithelium (Hartung et al. 1970; Melnick 1992; Melnick et al. 1994a; Smyth et al. 1951) all suggest that DEA affects the stability and function of cell membranes. An appropriate measure of dose and dose duration is needed so that equivalent "challenges" to target organs may be calculated for risk assessment. This may be accomplished using a physiological pharmacokinetic model to extrapolate data across different routes of exposure (i.e., inhalation, dermal, and oral) and between species (mouse, rat, and human). DEA was not genotoxic in the studies reviewed but was materno- and fetotoxic in a range-finding developmental toxicity study (EHRT 1990) at dose levels of 200 mg/kg and above. No materno- or fetotoxicity was observed at 50 mg/kg from day 6 to day 15 of gestation.

TEA is distinctly different from either MEA or DEA in that it is not metabolized to other compounds or incorporated into phospholipids. It is readily eliminated from body tissues and excreted in urine. TEA is the least basic compound of the three alkanolamines studied, with a pH$_a$ of 7.76. The histopathological changes produced by high doses of TEA (hyperplasia of the tubular epithelium, renal and hepatic adenomas, cloudy swelling and fatty changes in the liver, cloudy swelling of the convoluted tubules and loop of Henle) may be related to effects on structure, stability, and function of biomembranes during transport across lipid bilayers (Hejtmancik et al. 1985b, 1995; Kindsvatter 1940; Maekawa et al. 1986; NTP 1994). The presence of a *Helicobacter* species (*H. hepaticus*) in the livers of mice confounded the NTP dermal bioassay (Hejtmancik et al. 1995; NTP 1994). The effects observed in liver appear to be caused by *H. hepaticus* and not by TEA. The amount of TEA administered in mouse drinking water studies (up to 3000 mg kg^{-1} d^{-1}; Konishi et al. 1992; Maekawa et al. 1986) was extremely large compared to the amount of MEA/choline (10 mg kg^{-1} d^{-1}) required by animals to maintain lipid bilayers and remove lipids from liver. The histopathological changes observed in the clearance organs of rats (kidneys) and mice (liver) may also be related to nutritional deficiencies (i.e., choline, amino acids, etc.) brought on by high dietary levels and repeated dermal applications of TEA (for 30 d or more). According to Huber and Kidd (1990), dietary egg yolk phospholipid supplements show promise for alleviating the aging syndrome in animals associated with dysfunction of body tissues, while a choline–methionine diet prevents rats from developing hepatocarcinoma (Zeisel 1989). TEA per se was neither mutagenic, embryotoxic, nor teratogenic in published genotoxicity and reproduction studies.

The principal route of exposure to MEA, DEA, or TEA is through skin. Rat skin permeation studies (*in vivo* and *in vitro*) indicated that MEA, DEA, and TEA, when applied in water, penetrate skin at the rate of 2.9×10^{-3}, 4.36×10^{-3}, and 18×10^{-3} cm/hr, respectively. The penetration rate (k_p, cm/hr) appears to be inversely related to their respective pH_a values (9.3, 8.8, and 7.76) or to the number of hydroxyethyl groups (1, 2, or 3). Repeated applications of DEA to skin resulted in a 1.5- to 3.0-fold increase in absorption ($\mu g \, cm^{-2} \, hr^{-1}$) (Waechter et al. 1995). This increase in percutaneous absorption was less than the 450-fold increase produced using low to high concentrations of DEA ($0.2–20 \, mg/cm^2$) in single applications (Waechter et al. 1995). Repeated skin application of MEA, DEA, and TEA produced vacuolation of epithelial cells, thickening of the epithelium, inflammation, and necrosis (Hejtmancik et al. 1985c, 1987c; Melnick et al. 1988; Weeks et al. 1960). Workers exposed acutely or repeatedly to 20 mg/cm^2 or more of MEA, DEA, and TEA should wear protective clothing to prevent skin irritation and dermal absorption. The skin of workers repeatedly exposed to low concentrations ($0.2 \, mg/cm^2$) should be monitored routinely for signs of dermal irritation. Workers should frequently wash exposed skin and shower at the end of the workday.

Inhalation of vapor and aerosols constitutes a secondary route of exposure to MEA, DEA, and TEA in the workplace. Repeated whole-animal vapor/aerosol exposures resulted in significant dermal exposure and histopathological changes in skin. Respiratory tract changes associated with exposure to 5–15 ppm MEA consisted of nasal epithelium erosion and plasma cell infiltration in the dog, lymphocytic infiltration and general congestion of the lungs of the rat and guinea pig, and focal hemorrhages in dogs (Weeks et al. 1960). Limited respiratory data are available on DEA. High concentrations of DEA vapor/aerosols (200/1400 ppm, respectively) produced respiratory difficulty and death in rats (Hartung et al. 1970), while inhalation of TEA aerosols (0, 125, 250, 500, or 2000 mg/m^3) produced no lesions in the nasal mucosa or pulmonary tissues (Mosberg et al. 1985a). The observed effects appear to be directly related to the pH_a of the individual alkanolamines. Respiratory protection is recommended for MEA and DEA.

Physiological pharmacokinetic models are needed for the alkanolamines to facilitate route-to-route and animal-to-human extrapolations. The repeated dermal and inhalation dose responses (histopathology) were largely the result of multiroute exposures (i.e., dermal studies resulted in oral exposure, and inhalation studies resulted in dermal exposure). Repeated dose inhalation studies should have been conducted using nose-only inhalation chambers, while oral studies (drinking water or gavage) should have been conducted in place of the dermal studies to obtain systemic data.

Pharmacokinetic data should have been used for route-to-route dose extrapolation.

Summary

The chemistry, biochemistry, toxicity, and industrial use of monoethanolamine (MEA), diethanolamine (DEA), and triethanolamine (TEA) are reviewed. The dual function groups, amino and hydroxyl, make them useful in cutting fluids and as intermediates in the production of surfactants, soaps, salts, corrosion control inhibitors, and in pharmaceutical and miscellaneous applications. In 1995, the annual U.S. production capacity for ethanolamines was 447,727 metric tons.

The principal route of exposure is through skin, with some exposure occurring by inhalation of vapor and aerosols. MEA, DEA, and TEA in water penetrate rat skin at the rate of 2.9×10^{-3}, 4.36×10^{-3}, and 18×10^{-3} cm/hr, respectively. MEA, DEA, and TEA are water-soluble ammonia derivatives, with pHs of 9–11 in water and pH_a values of 9.3, 8.8, and 7.7, respectively. They are irritating to the skin, eyes, and respiratory tract, with MEA being the worst irritant, followed by DEA and TEA. The acute oral LD_{50}s are 2.74 g/kg for MEA, 1.82 g/kg for DEA, and 2.34 g/kg for TEA (of bw), with most deaths occurring within 4 d of administration.

MEA is present in nature as a nitrogenous base in phospholipids. These lipids, composed of glycerol, two fatty acid esters, phosphoric acid, and MEA, are the building blocks of biomembranes in animals. MEA is methylated to form choline, another important nitrogenous base in phospholipids and an essential vitamin. The rat dietary choline requirement is 10 mg kg^{-1} d^{-1}; 30-d oral administration of MEA (160–2670 mg kg^{-1} d^{-1}) to rats produced "altered" liver and kidney weights in animals ingesting 640 mg kg^{-1} d^{-1} or greater. Death occurred at dosages of 1280 mg kg^{-1} d^{-1}. No treatment-related effects were noted in dogs administered as much as 22 mg kg^{-1} d^{-1} for 2 yr.

DEA is not metabolized or readily eliminated from the liver or kidneys. At high tissue concentrations, DEA substitutes for MEA in phospholipids and is methylated to form phospholipids composed of N-methyl and N, N-dimethyl DEA. Dietary intake of DEA by rats for 13 wk at levels greater than 90 mg kg^{-1} d^{-1} resulted in degenerative changes in renal tubular epithelial cells and fatty degeneration of the liver. Similar effects were noted in drinking water studies. The findings are believed to be due to alterations in the structure and function of biomembranes brought about by the incorporation of DEA and methylated DEA in headgroups.

TEA is not metabolized in the liver or incorporated into phospholipids. TEA, however, is readily eliminated in urine. Repeated oral administration to rats (7 d/wk, 24 wk) at dose levels up to and including 1600 mg kg^{-1} d^{-1} produced histopathological changes restricted to kidney and liver. Lesions in the liver consisted of cloudy swelling and occasional fatty changes, while cloudy swelling of the convoluted tubules and loop of Henle were observed in kidneys. Chronic administration (2 yr) of TEA in drinking water (0, 1%, or 2% w/v; 525 and 1100 mg kg^{-1} d^{-1} in males and 910 and 1970 mg kg^{-1} d^{-1} in females) depressed body and kidney weights in F-344 rats. Histopathological findings consisted of an "acceleration of so-called chronic nephropathy" commonly found in the kidneys of aging F-344 rats. In B$_6$C$_3$F$_1$ mice, chronic administration of TEA in drinking water (0, 1%, or 2%) produced no significant change in terminal body weights between treated and control animals or gross pathological changes. TEA was not considered to be carcinogenic. Systemic effects in rats chronically administered TEA dermally (0, 32, 64, or 125 mg kg^{-1} d^{-1} in males; 0, 63, 125, or 250 mg kg^{-1} d^{-1} in females) 5 d/wk for 2 yr were primarily limited to hyperplasia of renal tubular epithelium and small microscopic adenomas. In a companion mouse dermal study, the most significant change was associated with nonneoplastic changes in livers of male mice consistent with chronic bacterial hepatitis.

References

Aldrich Library of Infrared Spectra (ALIRS) (1981) ed III. Aldrich Chemical Co., Milwaukee, WI.

Annau E, Manginelli A (1950) Alkaline phosphatase activity and nuclear changes in the liver induced by diethanolamine. Nature 166:816–817.

Annau E, Manginelli A, Roth A (1950) Biochemical and biological effects of diethanolamine. Glycogen and lipid content of the liver of mice receiving simple alkanolamines. Nature 166:815–816.

Ansell GB, Spanner S (1966) The incorporation of [2-^{14}C]ethanolamine and [methyl-^{14}C]choline into brain phospholipids in vivo and in vitro. Proc Biochem Soc 100:50.

Araksyan AM (1960) Effect of di- and triethanolamine on the oxidative processes of isolated white rat liver tissue. Erevansk Zootekln Vet Inst 24:47.

Artom C, Cornatzer WE, Crowder M (1949) The action of an analogue of ethanolamine (diethanolamine) on the formation of liver phospholipids. J Biol Chem 180:495–503.

Arutyunyan RM, Zalinyan RM, Mugnetsyan EG, Gukasyan LA (1987) Mutagenic action of latex polymerization stabilizers in different test systems. Tsitol Genet 21:450–456.

Babior BM (1969) Mechanism of action of ethanolamine deaminase. I. Studies with isotopic hydrogen and oxygen. J Biol Chem 244:449–456.

Barbee SJ, Hartung R (1979a) The effect of diethanolamine on hepatic and renal phospholipid metabolism in the rat. Toxicol Appl Pharmacol 47:421–430.

Barbee SJ, Hartung R (1979b) Diethanolamine-induced alteration of hepatic mitochondrial function and structure. Toxicol Appl Pharmacol 47:431–440.

Barthelmess A, Elkabarity A (1962) Chemisch Indirzierte Multipolare Mitosen III. Protoplasma LIV:455–475.

Battelle (1988a) Mating trial dermal study of triethanolamine in Fischer-344 rats. Battelle Columbus Laboratories, Columbus, OH.

Battelle (1988b) Mating trial dermal study of triethanolamine in Swiss CD-1 mice. Battelle Columbus Laboratories, Columbus, OH.

Beyer KH, Bergfeld WF, Berndt WO, Boutwell RK, Carlton WW, Hoffmann DK, Schroeter AL (1983) Final report on the safety assessment of triethanolamine, diethanolamine, and monoethanolamine. J Am Coll Toxicol 2:183-235.

BIBRA (1990) Triethanolamine toxicity profile, 2nd ed. British Industrial Biological Research Association, Carshalton Surrey, U.K.

BIBRA (1993a) Ethanolamine toxicity profile, 2nd ed. British Industrial Biological Research Association, Carshalton Surrey, U.K.

BIBRA (1993b) Diethanolamine toxicity profile, 2nd ed. British Industrial Biological Research Association, Carshalton Surrey, U.K.

Binks SP, Smillie MV, Glass DC, Flethcher AC, Shackleton S, Robertson AS, Levy LS, Chipman JK (1992) Occupational exposure limits. Criteria document for ethanolamine. Commission of the European Communities, Luxembourg.

Blum K, Huizenga CG, Ryback RS, Johnson DK, Geller I (1972) Toxicology of diethanolamine in mice. Toxicol Appl Pharmacol 22:175-185.

Bose S, Naskar SK (1975) Effect of dimethyl sulfoxide, ethylene glycol, hydroxyl-amine and triethanolamine in M_1 generation in cluster bean. Bull Bot Soc Bengal 29:49-52.

Brabec MJ, Gray RH, Bernstein IA (1974) Restoration of hepatic mitochondria during recovery from carbon tetrachloride intoxication. Biochem Pharmacol 23: 3227-3238.

Bronaugh RL, Stewart RF, Congdon ER, Giles AL (1982) Methods for *in vitro* percutaneous absorption. I. Comparison with *in vivo* results. Toxicol Appl Pharmacol 62:474-480.

Bronaugh RL, Steward RF, Simon M (1986) Methods for *in vitro* percutaneous absorption studies. VII: Use of excised human skin. J Pharm Sci 75(11):1094.

Burnett C, Goldenthal EI, Harris SB, Wazeter FX, Strausburg J, Kapp R, Voelker R (1976) Teratology and percutaneous toxicity studies on hair dyes. J Toxicol Environ Health 1:1027-1040.

Byard JL, Koepki UCH, Abraham R, Goldberg L, Coulston F (1975) Biochemical changes in the liver of mice fed Mirex. Toxicol Appl Pharmacol 33:70-77.

Chernoff N, Kavlock RJ (1982) An *in vivo* teratology screen utilizing pregnant mice. J Toxicol Environ Health 10:541-550.

Chojnacki T, Korzybski T (1963) The transfer of the phosphoric ester of *N,N*-diethyl-aminoethanol from its cytidylyl derivative into phospholipids. Acta Biochim Pol. 10:233-241.

Coblentz Collection (COB) (1969-1970) Joint Committee on Atomic and Molecular Physical Data, Evaluated IR Spectra, BioRad Laboratories, Stadtler Division, Philadelphia, PA.

Cosmetic Ingredient Review (CIR) Expert Panel (1983) Final report on the safety assessment of triethanolamine, diethanolamine, and monoethanolamine. Cosmetic Fragrance Ingredient Review. J Am Coll Toxicol 2:183-235.

Davidson A, Milwidsky B (1968) Synthetic detergents, 4th ed. CRC Press, Cleveland, OH, pp 131-136.

Dean JA (ed) (1992) Lange's Handbook of Chemistry, 14th ed. McGraw-Hill, New York.

Dean DJ, Brooks TM, Hodson-Walker G, Hutson DH (1985) Genetic toxicology testing of 41 industrial chemicals. Mutat Res 153:57–77.

DePass LR, Fowler EH, Leung H-W (1995) Subchronic dermal toxicity study of triethanolamine in C3H/HeJ mice. Food Chem Toxicol 33:675–680.

Douglas ML, Kabacoff BL, Anderson GA, Cheng MC (1978) The chemistry of nitrosamine formation, inhibition and destruction. J Soc Cosmet Chem 29:581–606.

Dow Chemical Company (1962) Gas conditioning fact book. The Dow Chemical Company, Midland, MI.

Dow Chemical Company (1988) The alkanolamines handbook. The Dow Chemical Company, Midland, MI.

Eagle H (1959) Amino acid metabolism in mammalian cell culture. Science 130:432.

ECETOC (1990) Human exposure to N-nitrosamines, their effects, and a risk assessment for N-nitrosodiethanolamine in personal care products. Tech Rep 41. European Chemical Industry.

EHRT (1987) Screening of priority chemicals for reproductive hazards. Monoethanolamine, diethanolamine and triethanolamine. ETOX-85-1002. Environmental Health Research & Testing, Cincinnati, OH.

EHRT (1990) Range-finding studies: developmental toxicity of diethanolamine when administered via gavage in CD Sprague–Dawley rats. NTP-89-RF/DT-002. Environmental Health Research & Testing, Lexington, KY.

Estabrook RW (1967) Mitochondrial respiratory control and the polarographic measurements of ADP/O ratios. In: Colowick S, Kaplan NO (eds) Methods of enzymology, vol. X. Academic Press, New York, pp 41–47.

Foster G (1971) Studies of the acute and subacute toxicological responses to diethanolamine in the rat. Doctoral dissertation, University of Michigan, Ann Arbor, MI.

Galloway SM, Armstrong MJ, Reuben C, Colman S, Brown B, Cannon C, Bloom AD, Nakamura F, Ahmed M, Duk S, Rimpo J, Margolin BH, Resnick MA, Anderson B, Zeiger E (1987) Chromosomal aberrations and sister chromatid exchanges in Chinese hamster ovary cells: evaluations of 108 chemicals. Environ Mol Mutagen 10:1–175.

Gibaldi M, Perrier D (1982) Pharmacokinetics, 2nd ed. Dekker, New York.

Hartung R, Rigas LK, Cornish HH (1970) Acute and chronic toxicity of diethanolamine (abstract). Toxicol Appl Pharmacol 17(1):308.

Haworth S, Lawlor T, Mortelmans K, Speck W, Zeiger E (1983) Salmonella mutagenicity test results for 250 chemicals. Environ Mut (Suppl) 1:3–142.

Hedenstedt A (1978) Mutagenicity screening of industrial chemicals: seven aliphatic amines and one amide tested in the Salmonella/microsomal assay (abstract). Mutat Res 53:198–199.

Hejtmancik M, Mezza L, Peters AC, Athey PM (1985a) The repeated dose dosed water study of triethanolamine (CAS No. 102-71-6) in Fischer-344 rats. Battelle Columbus Division Laboratories, Columbus, OH.

Hejtmancik M, Mezza L, Peters AC, Athey PM (1985b) The repeated dose dosed water study of triethanolamine (CAS No. 102-71-6) in $B_6C_3F_1$ mice. Battelle Columbus Division Laboratories, Columbus, OH.

Hejtmancik M, Mezza L, Peters AC, Athey PM (1985c) The dermal repeated dose study of triethanolamine (CAS No. 102-71-6) in Fischer-344 rats. Battelle Columbus Division Laboratories, Columbus, OH.

Hejtmancik M, Mezza L, Peters AC, Athey PM (1985d) The dermal repeated dose study of triethanolamine (CAS No. 102-71-6) in $B_6C_3F_1$ mice. Battelle Columbus Division Laboratories, Columbus, OH.

Hejtmancik M, Athey PM, Ryan MJ, Peters AC (1987a) The repeated dose dosed water study of diethanolamine (CAS No. 111-42-2) in Fischer-344 rats. Battelle Columbus Division Laboratories, Columbus, OH.

Hejtmancik M, Athey PM, Persing RL, Peters AC (1987b) The repeated dose dosed water study of diethanolamine (CAS No. 111-42-2) in $B_6C_3F_1$ mice. Battelle Columbus Division Laboratories, Columbus, OH.

Hejtmancik M, Johnson JD, Ryan MJ, Athey PM, Peters AC (1987c) The repeated dose dermal study of diethanolamine (CAS No. 111-42-2) in Fischer-344 rats. Battelle Columbus Division Laboratories, Columbus, OH.

Hejtmancik M, Johnson JD, Persing RL, Athey PM, Peters AC (1987d) The repeated dose dermal study of diethanolamine (CAS No. 111-42-2) in $B_6C_3F_1$ mice. Battelle Columbus Division Laboratories, Columbus, OH.

Hejtmancik M, Mezza L, Athey PM, Peters AC (1987e) The prechronic studies of triethanolamine (CAS No. 120-71-6) in Fischer-344 rats. Battelle Columbus Division Laboratories, Columbus, OH.

Hejtmancik M, Mezza L, Athey PM, Peters AC (1987f) The prechronic studies of triethanolamine (CAS No. 120-71-6) in $B_6C_3F_1$ mice. Battelle Columbus Division Laboratories, Columbus, OH.

Hejtmancik M, Mezza L, Peters AC (1988) Prechronic dosed water study of diethanolamine (CAS No.111-42-2) in Fischer-344 rats. Battelle Columbus Division Laboratories, Columbus, OH.

Hejtmancik M, Toft JD, Persing RL, Melnick RL (1995) Two-year dermal study of triethanolamine in F344 rats and $B_6C_3F_1$ mice (abstract). Toxocologist 15:202.

Hellwig J, Hildebrand B (1994) Study of the pre-, peri-, postnatal toxicity of monoethanolamine pure in rats after oral administration (gavage). Project report 60R0351/91062, BASF Aktiengesellschaft, Ludwigshafen, Germany.

Holland JM, Kao JY, Whitaker MJ (1984) A multi-sample apparatus for kinetic evaluation of skin penetration in vitro: the influence of variability and metabolic status of the skin. Toxicol Appl Pharmacol 72:272–280.

Horrocks LA (1969) Metabolism of the ethanolamine phosphoglycerides of mouse brain myelin and microsomes. J Neurochem 16:13–18.

Hoshino H, Tanooka H (1978) Carcinogenicity of triethanolamine in mice and its mutagenicity after reaction with sodium nitrite in bacteria. Can Res 38:3918–3921.

Howe-Grant M (ed) (1992) Encycopedia of Chemical Technology, 4th ed. Vol. 2, Alkanolamines to Antibiotics (glycopeptides). Wiley, New York.

Huber W, Kidd PM (1990) Dietary egg yolk-derived phospholipids: rationale for their benefits in syndromes of senescence, drug withdrawal, and AIDS. In: Hanin I, Pepeu G (eds) Phospholipids: Biochemical, Pharmaceutical, and Analytical Considerations. Plenum Press, New York, pp 241–255.

Hui SW, Viswanathan R, Zasadzinski JA, Israelachvill JN (1995) The structure and stability of phospholipid bilayers by atomic force microscopy. Biophys J 68:171–178.

Inai K, Aoki Y, Akamizu H, Eto R, Nishida T, Tokuoka S (1979) Tumorigenicity study of butyl and isobutyl p-hydroxybenzoates administered orally to mice. Food Chem Toxicol 23:575–578.

Inai K, Aoki Y, Tokuoka S (1985) Chronic toxicity of sodium nitrite in mice, with reference to its tumorigenicity. Gann 70:203–208.

Inoue K, Sunakawa T, Okamoto K, Tanaka Y (1982) Mutagenicity tests and *in vitro* transformation assays on triethanolamine. Mutat Res 101:305-313.

Jungalwala FB, Dawson RMC (1970a) Phospholipid synthesis and exchange in isolated liver cells. Biochem J 117:481-490.

Jungalwala FB, Dawson RMC (1970b) The origin of mitochondrial phosphatidyl choline within the liver cell. Eur J Biochem 12:399-402.

Jungerman E, Tabor D (1967) Nonionic surfactants. Dekker, New York, pp 226-239.

Kao JY, Hall J, Holland JM (1983) Quantitation of cutaneous toxicity: an *in vitro* approach using skin organ culture. Toxicol Appl Pharmacol 68:206-217.

Kartner N, Ling V (1989) Multidrug resistance in cancer. Sci Am 260:44-51.

Kindsvatter VH (1940) Acute and chronic toxicity of triethanolamine. J Ind Hyg Toxicol 22:206-212.

Konishi Y, Denda A, Kazuhiko U, Yohko E, Hitoshi U, Yokose Y, Shiraiwa K, Tsutsumi M (1992) Chronic toxicity carcinogenicity studies of triethanolamine in $B_6C_3F_1$ mice. Fundam Appl Toxicol 18:25-29.

Korhonen A, Hemminski K, Vainio H (1983) Embryotoxicity of sixteen industrial amines to the chicken embryo. J Appl Toxicol 3:112-117.

Kostrodymova GM, Voronin VM, Kostrodymova NN (1976) The toxicity and the possibility of cancerogenic and cocancerogenic properties of triehthanolamines [Engl transl]. Gig Sanit 3:20-25.

Kuo JK (1985) Phospholipids and Cellular Regulation. CRC Press, Boca Raton, FL.

Liberacki AB, Neeper-Bradley TL, Breslin WJ, Zielke GJ (1996) Evaluation of the developmental toxicity of dermally applied monoethanolamine in rats and rabbits. Fundam Appl Toxicol 31:117-123.

Lijinsky W, Keefer L, Conrad E, Van de Bogart R (1972) Nitrosation of tertiary amines and some biologic implications. J Natl Cancer Inst 49:1239-1249.

Lijinsky W, Kovatch RM (1985) Induction of liver tumors in rats by nitrosodiethanolamine at low doses. Carcinogenesis (Oxford) 6:1679-1681.

Litton Bionetics (1982) Evaluation of triethanolamine no. 80/475 in the primary rat hepatocyte unscheduled DNA synthesis assay. Final report. In: Triethanolamin, Toxikologische Bewertung. Rep 57. Berufsgenossenschaft der Chemischen Industrie, Heidelberg.

Loveday KS, Lugo MH, Resnick MA, Anderson BE, Zeiger E (1989) Chromosome aberration and sister chromatid exchanges in Chinese hamster ovary cells *in vitro*: II. Results with 20 chemicals. Environ Mol Mutagen 13:60-94.

Maekawa A, Onodera H, Tanigawa H, Furuta K, Kanno J, Matsuoka C, Oglu T, Hayashi Y (1986) Lack of carcinogenicity of triethanolamine in F344 rats. J Toxicol Environ Health 19:345-357.

Mankes RF (1986) Studies on the embroyopathic effects of ethanolamine in Long-Evans rats: preferential embryopathy in pups contiguous with male siblings in utero. Teratogen Carcinogen Mutagen 6:403-417.

Mathews JM, Jeffcoat AR (1991) Absorption and disposition of diethanolamine (DEA) in rats and mice after oral, dermal, and intravenous administration. RTI/3662/00-12P. Research Triangle Institute, NC.

Mathews JM, Garner CE, Mathews HB (1993) Metabolism and bioaccumulation of diethanolamine (DEA) in rats (abstract). Toxicologist 13:1578.

Mathews JM, Garner CE, Mathews HB (1995) Metabolism, bioaccumulation, and incorporation of diethanolamine into phospholipids. Chem Res Toxicol 8:625–633.

McMurray WC, Dawson RMC (1969) Phospholipid exchange reactions within the liver cell. Biochem J 112:91–108.

Melnick R, Hejtmancik M, Mezza L, Ryan M, Persing RL, Peters A (1988) Comparative effects of triethanolamine (TEA) and diethanolamine (DEA) in short-term dermal studies (abstract). Toxicologist 8:505.

Melnick RL (1992) NTP technical report on toxicity studies of diethanolamine administered topically and in drinking water to F344/N rats and $B_6C_3F_1$ mice. Publ 92-3343. National Institutes of Health, Bethesda, MD.

Melnick RL, Mahler J, Bucher JR, Thompson M, Hejtmancik M, Ryan MJ, Mezza LE (1994a) Toxicity of diethanolamine. 1. Drinking water and topical application exposures in F344 Rats. J Appl Toxicol 14:1–9.

Melnick RL, Mahler J, Bucher JR, Hejtmancik M, Singer A, Persing RL (1994b) Toxicity of diethanolamine. 2. Drinking water and topical application exposures in $B_6C_3F_1$ mice. J Appl Toxicol 14:11–19.

Mervish SS (1975) Formation of N-nitroso compounds: chemistry, kinetics, and in vivo occurrences. Toxicol Appl Pharmacol 31:325–351.

Michael UF, Cookson SL, Chavez R, Pardo V (1975) Renal function in the choline deficient rat. Proc Soc Exp Biol Med 150:672–676.

Morin RJ (1969) In vitro inhibition by metabolic antagonists of incorporation of ^{32}P-phosphate into the major phospholipids of swine coronary and pulmonary arteries. J Atheroscler Res 10:283–289.

Mortelmans K, Haworth S, Lawlor T, Speck W, Tainer B, Zeiger E (1986) Salmonella mutagenicity tests: II. Results from the testing of 270 chemicals. Environ Mutagen 8:1–119.

Mosberg A, McNeill D, Hejtmancik M, Persing RL, Peters A (1985a) The repeated dose inhalation study of triethanolamine (CAS No. 102-71-6) in Fischer-344 rats. Battelle Columbus Division Laboratories, Columbus, OH.

Mosberg A, McNeill D, Hejtmancik M, Persing RL, Peters A (1985b) The repeated dose inhalation study of triethanolamine (CAS No. 102-71-6) in $B_6C_3F_1$ mice. Battelle Columbus Division Laboratories, Columbus, OH.

Myhr BC, Bowers LR, Caspary WJ (1986) Results from the testing of coded chemicals in the L5178Y TK + / − mouse lymphoma mutagenesis assay (abstract). Environ Mutagen 7(Suppl 3):58.

Neeper-Bradley TL (1992) Developmental toxicity evaluation of diethanolamine applied cutaneously to CD rats and New Zealand white rabbits. Project 54-563. Union Carbide Bushy Run Research Center, Export, PA.

Neeper-Bradley TL, Kubena MF (1993) Diethanolamine (DEA): developmental toxicity study of cutaneous administration to New Zealand White rabbits. Project 91N0136. Union Carbide Bushy Run Research Center, Export, PA.

NIST/EPA/MSDC (1978–1980) Mass Spectral database. U.S. National Institutes of Standards and Technology (NIST), Office of Standard Reference Data, Gaithersburg, MD.

NRC (1981) Selected aliphatic amines and related compounds: an assessment of the biological and environmental effects. PB83-133066. Committee on Amines, Board on Toxicology and Environmental Health Hazards, Assembly of Life Sciences, National Research Council, Washington, DC.

NTP (National Toxicology Program) (1992) Toxicity studies of diethanolamine administered topically and in drinking water to F344/N rats and $B_6C_3F_1$ mice. TRS #20, National Toxicology Program. Publ 92-3343. National Institutes of Health (NIH), Bethesda, MD.

NTP (1994) Technical report on the toxicology and carcinogenesis studies of triethanolamine (CAS No. 102-71-6) in F344/N rats and $B_6C_3F_1$ mice (dermal studies). NTP TR #449. Publ 94-3365 NIH, Bethesda, MD.

OCC (1995) (Occidental Chemical Corporation) Technical service reference manual, ethanolamines. Occidental Chemical Corp., Niagara Falls, NY.

Pathak MA, Long SD, Warren AJ (1982) Mutagenicity studies of ultraviolet-absorbing sunscreens and dihydroxyacetone (abstract). Clin Res 30:265.

Polish Patent (1987) 140,720. J. Perger and co-workers to Nadorzanskie Zaklady Przemyslu Organiczanego "Organika-Rokita."

Preussman R, Habs M, Schmahl D, Eisenbrand G (1981) Urinary excretion of N-nitrosodiethanolamine in rats following its epicutaneous and intratracheal administration and its formation in vivo following skin application of diethanolamine. Cancer Lett 13:227-232.

Preussman R, Habs M, Schmahl D (1982) Carcinogenicity of N-nitrosodiethanolamine in rats at five different dose levels. Cancer Res 42:5167-5171.

Roehm DC, Hooberry JJ (1959) Uses of the N-alkanolamines in treatment of diseases of metabolism: Gallogen, Deaner, diethanolamine (DEA) and glucosamine HCl. Clin Res 7:148.

Roehm DC (1973) Production of fibrinolysis, anticoagulation and hypocholesteremia by diethanolamine (DEA). Clin Res 7:148.

Roehm DC (1975) Method of controlling lipids in the bloodstream. U.S. Patent 3,892,865.

Sanger SL, Nicholson GL (1972) The fluid mosaic model of the structure of cell membranes. Science 175:720-731.

Schneider WC (1963) Intracellular distribution of enzymes. XIII. Enzymatic synthesis of deoxycytidine diphosphatecholine and lecithin in rat liver. J Biol Chem 238:3572-3578.

Seilkop SK (1995) The effect of body weight on tumor incidence and carcinogenicity testing in $B_6C_3F_1$ mice and F344 rats. Fundam Appl Toxicol 24:247-259.

Smyth HF, Carpenter CP, Weil CS (1951) Range-finding toxicology data: List IV A.M.A. Arch Ind Hyg Occup Med 4:119-122.

Sorsa M, Pyg L, Salomaa C, Nylund L, Yager JW (1988) Biological and environmental monitoring of occupational exposure to cyclophosphamide in industry and hospitals. Mutat Res 204:465-479.

Sprinson DB, Weliky I (1969) The conversion of ethanolamine to acetate in mammalian tissues. Biochem Biophys Res Commun 36:866-870.

SRI International (1995) 1995 Directory of Chemical Producers. SRI International, Menlo Park, CA, p 583.

Sun JD, Beskitt JL, Tallant MJ, Frantz SW (1996) In vitro skin penetration of monoethanolamine and diethanolamine using excised skin from rats, mice, rabbits and humans. J Toxicol Cutan Ocul Toxicol 15(2):131-146.

Sundler R (1973) Biosynthesis of rat liver phosphatidylethanolamines from intraportally injected ethanolamine. Biochim Biophys Acta 306:218-226.

Sundlof SF, Mayhew IG (1983) A neuroparalytic syndrome associated with an oral flea repellant containing diethanolamine. Vet Hum Toxicol 25:247-249.

Taylor RJ, Richardson KE (1967) Ethanolamine metabolism in the rat. Proc Soc Exp Biol Med 124:247–252.

Tennant RW, French JE, Spalding JW (1995) Identifying chemical carcinogens and assessing potential risk in short-term bioassays using transgenic mouse models. Environ Health Perspect 103:942–950.

Timofievskaya LA (1962) Toxicologic characteristic of aminoethanol. Toksikologiya Novykh Promysch Khimicheskike Veshchestv 4:81–91.

Tinoco J, Sheehan G, Hopkins S, Lyman RL (1970) Incorporation of (1,2-^{14}C)-ethanolamine into subfractions of rat liver phosphatiylethanolamines and phosphatidylcholines. Lipids 5:412–416.

Treon JF, Cleveland FP, Stemmer KL, Cappel J, Shaffer F, Largent EE (1957) The toxicity of monoethanolamine in air. Dettering Laboratory, University of Cincinnati, Cincinnati, OH.

TRGS (1988) Technische Regeln fur Gefahrstoffe: Nitrosamine. No. 552. Carl Heymanns Verlag, Koeln.

UCC (Union Carbide Corporation) (1982) Monoethanolamine dilutions: Department of Transportation skin irritancy test. Project 45-137. Union Carbide Corporation Bushy Run Research Center, Export, PA.

UCC (1988a) Monoethanolamine: acute toxicity and primary irritancy studies. Project 51-86. Union Carbide Corporation Bushy Run Research Center, Export, PA.

UCC (1988b) Diethanolamine: acute toxicity and primary irritancy studies. Project 51-95. Union Carbide Corporation Bushy Run Research Center, Export, PA.

UCC (1988c) Triethanolamine: acute toxicity and primary irritancy studies. Project 51-94. Union Carbide Corporation Bushy Run Research Center, Export, PA.

USSR (1989) Patent 1,484,850, L.P. Loseva to Ivanovo Chemical-Technological Institute, and Ivanovo Worsted Fabric Combine.

Waechter JM, Rick DL (1988) Triethanolamine: pharmacokinetics in C3H/HeJ mice and Fischer-344 rats following dermal administration. Dow Chemical Company, Midland, MI.

Waechter JM, Bormett GA, Stewart HS (1995) Diethanolamine: pharmacokinetics in Sprague–Dawley rats following dermal or intravenous administration. Dow Chemical Company, Midland, MI.

Wang D, Huang W-Q, Wang H-W (1988) Mutagenicity and carcinogenicity studies of homemade "rust-proof cutting fluid." Teratogen Carcinogen Mutagen 8:35–43.

Ward JM, Anver MR, Haines DC, Benveniste RE (1994a) Chronic active hepatitis in mice caused by *Helicobacter hepaticus*. Am J Pathol 145:959–968.

Ward JM, Fox JG, Anver MR, Haines DC, George CV, Collins MJ, Gorelick PL, Nagashima K, Gonda MA, Gilden RV, Tully JG, Russell RJ, Benveniste RE, Paster BJ, Dewhirst FE, Donovan JC, Anderson LM, Rice JM (1994b) Chronic active hepatitis and associated liver tumors in mice caused by a persistent bacterial infection with a novel *Helicobacter* species. J Natl Cancer Inst 86:1222–1227.

Wattiaux-De Coninck S, Wattiaux R (1971) Subcellular distribution of sulfite cytochrome *c* reductase in rat liver tissue. Eur J Biochem 19:552–556.

Weeks MH, Downing TO, Musselman NP, Carson TR, Groff WA (1960) The effects of continuous exposure of animals to ethanolamine vapor. Am Ind Hyg Assoc J 21:374–381.

Welch AD, Landau RL (1942) The arsenic analog of cholines as a component of lecithin in rats fed arsenocholine chloride. J Biol Chem 144:581–588.

Wells IC, Remy CN (1961) Inhibition of de novo choline biosynthesis by 2-amino-2-
    methyl-1-propanol. Arch Biochem 95:389–399.
Wernick T, Lanmam BM, Fraux JL (1975) Chronic toxicity, teratologic, and repro-
    duction studies with hair dyes. Toxicol Appl Pharmacol 32:45–460.
White S (1994) Protein Membrane Structure. Oxford University Press, Oxford.
Wilgram GF, Kennedy EP (1963) Intracellular distribution of some enzymes cata-
    lyzing reactions in the biosynthesis of complex lipids. J Biol Chem 238:2615–
    2619.
Wise EM, Elwyn D (1965) Rates of reactions involved in phosphatide syntheses in
    liver and small intestine of intact rats. J Biol Chem 240:1537–1548.
Yoon JS, Mason JM, Valencia R, Woodruff RC, Zimmering S (1985) Chemical
    mutagenesis testing in *Drosophila*. IV. Results of 45 coded compounds tested for
    the National Toxicology Program. Environ Mutagen 7:349–367.
York RG, Barnwell PL, Pierrera M, Schuler RL, Hardin BD (1988) Evaluation of
    twelve chemicals in a preliminary developmental toxicity test. Teratology 37:
    503–504.
Zeisel SH (1989) Phospholipids and choline deficiency. In: Hanin I, Pepeu G (eds)
    Proceedings of the 5th International Colloquium on Lecithin (Cannes, France).
    Phospholipids: Biochemical, Pharmaceutical, and Analytical Considerations.
    Plenum Press, New York, pp 219–231.

Manuscript received February 23, 1996; accepted June 13, 1996.

Environmental Assessment of the Alkanolamines

John W. Davis*,‡ and Constance L. Carpenter†

Contents

I. Introduction

The alkanolamine product family consists of the ethanol-, isopropanol-, and butanol-substituted amines. The alkanolamines are bifunctional molecules having both amino and alcohol functional groups. As a result, they undergo a wide variety of reactions common to amines and alcohols. Because of these physicochemical characteristics, the alkanolamines are used in a wide variety of applications, including surfactants, cosmetics, toiletry products, metalworking fluids, textile chemicals, gas conditioning chemicals, agricultural chemical intermediates, and cement grinding aids. Due to this wide range of uses and applications, there is a need to understand the fate and effects of these compounds in the environment.

This review provides a summary of the current information available on the environmental fate and aquatic toxicology of the alkanolamines product family. The following alkanolamines are included in this review: monoethanolamine (MEA), diethanolamine (DEA), triethanolamine (TEA), monoisopropanolamine (MIPA), diisopropanolamine (DIPA), triisopropanolamine (TIPA), monobutanolamine (1-amino-2-butanol) (MBA), di-

---

*Environmental Chemistry Research Laboratory, Health and Environmental Sciences, The Dow Chemical Company, Midland, MI 48674, U.S.A.

†DowBrands, The Dow Chemical Company, Urbana, OH 43078, U.S.A.

‡Corresponding author.

Reviews of Environmental Contamination and Toxicology, Vol. 149.

butanolamine (DBA), and tributanolamine (TBA) (Table 1). Information in this assessment includes a review of selected physical characteristics and the potential for degradation of the alkanolamines in the atmosphere, soil, surface water, and groundwater, as well as the potential for wastewater treatability. Information on the aquatic toxicology and bioconcentration potential of these compounds was also evaluated.

II. Physical Properties

Selected physical properties of the ethanol-, isopropanol-, and butanolamines are provided in Tables 2–4. This information is not meant to be a comprehensive listing of the physical properties of alkanolamines but was included in evaluating their environmental fate. A unit world model developed by Mackay and Paterson (1981) was used to estimate the distribution of the alkanolamines upon entry into the environment. This model incorporates a variety of physicochemical characteristics in order to assess the relative concentration of a compound in several environmental compartments. Variables required in the model included temperature (T), molecular weight (MW), octanol–water partition coefficient (K_{ow}), water solubility (S), and vapor pressure (P_{vp}). Several of the physicochemical characteristics for the alkanolamines required by this fugacity model have not been determined experimentally. In order to conduct the model simulations, these parameters were estimated using a variety of established methods (see following).

Henry's Law constants for the ethanolamines, isopropanolamines, and butanolamines are presented in Tables 2–4. These constants ranged from 3.39×10^{-6} to 3.38×10^{-19} atm·m^3/mol. These constants relate the concentration of a compound between the air and aqueous phases and provide a relative indication of a chemical's volatility. The Henry's Law constants for highly volatile compounds are routinely $\geq 10^{-3}$ atm·m^3/mol, while the values for low volatility or nonvolatile compounds are usually $< 10^{-7}$ atm·m^3/mol (Lyman et al. 1982). Based upon these ranges, the ethanolamines, isopropanolamines, and butanolamines would be considered relatively nonvolatile.

The vapor pressure of a chemical has been shown to be an important parameter in evaluating its potential to evaporate from aqueous environments. Vapor pressures for the alkanolamines were estimated using Eq. (1) (Lyman et al. 1982).

A. Vapor Pressure

$$\ln P_{vp} = [\text{Æ}H_{vp} \times (T_b - C_2)^2 / \text{Æ}Z_b \times R \times (T_b)^2] \qquad (1)$$
$$\times [1/T_b - C_2) - 1/(T - T_b],$$

where $\text{Æ}H_{vp} = T_b \times K_F(8.75 + R \ln T_b)$, T_b is the boiling point in °K from the literature, T the temperature in °K, $C_2 = -18 + 0.19Tb$ (no

Table 1. The alkanolamines.

Name	IUPAC	CAS Number
Ethanolamines		
Monoethanolamine (MEA)	2-aminoethanol	141-43-5
Diethanolamine (DEA)	2,2'-iminodiethanol	111-42-2
Triethanolamine (TEA) 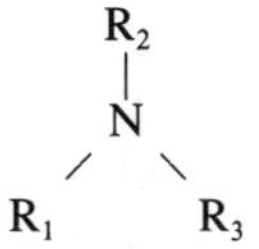	2,2',2"-nitrilotriethanol	102-71-6

$$R_2$$
$$|$$
$$N$$
$$/ \quad \backslash$$
$$R_1 \qquad R_3$$

MEA: R_1 = CH$_2$CH$_2$(OH); R_2 = H; R3 = H
DEA: R_1 = CH$_2$CH$_2$(OH); R_2 = CH$_2$CH$_2$(OH); R_3 = H
TEA: R_1 = CH$_2$CH$_2$(OH); R_2 = CH$_2$CH$_2$(OH); R_3 = CH$_2$CH$_2$(OH)

Isopropanolamines		
Monoisopropanolamine (MEA)	1-amino-2-propanol	78-96-6
Diisopropanolamine (DIPA)	1,1'-iminodi-2-propanol	110-97-4
Triisopropanolamine (TIPA) 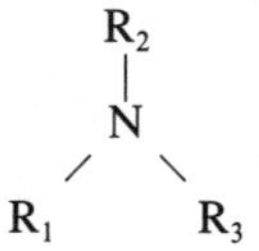	1,1',1"-nitrilotri-2-propanol	122-20-3

$$R_2$$
$$|$$
$$N$$
$$/ \quad \backslash$$
$$R_1 \qquad R_3$$

MIPA: R_1 = CH$_2$CH(OH)CH$_3$; R_2 = H; R_3 = H
DIPA: R_1 = CH$_2$CH(OH)CH$_3$; R_2 = CH$_2$CH(OH)CH$_3$; R_3 = H
TIPA: R_1 = CH$_2$CH(OH)CH$_3$; R_2 = CH$_2$CH(OH)CH$_3$; R_3 = CH$_2$CH(OH)CH$_3$

Butanolamines		
Monobutanolamine (MBA)	1-amino-2-butanol	13552-21-1
Dibutanolamine (DBA)	1,1'-iminodi-2-butanol	21838-75-5
Tributanolamine (TBA) 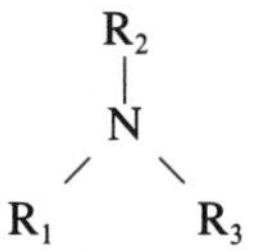	1,1',1"-nitrilotri-2-butanol	2421-02-5

$$R_2$$
$$|$$
$$N$$
$$/ \quad \backslash$$
$$R_1 \qquad R_3$$

MBA: R_1 = CH$_2$CH(OH)CH$_2$CH$_3$; R_2 = H; R_3 = H
DBA: R_1 = CH$_2$CH(OH)CH$_2$CH$_3$; R_2 = CH$_2$CH(OH)CH$_2$CH$_3$; R_3 = H
TBA: R_1 = CH$_2$CH(OH)CH$_2$CH$_3$; R_2 = CH$_2$CH(OH)CH$_2$CH$_3$; R_3 = CH$_2$CH(OH)CH$_2$CH$_3$

IUPAC, International Union of Pure and Applied Chemistry; CAS, Chemical Abstracts Service.

Table 2. Selected physical properties of the ethanolamines.

Property	Monoethanolamine	Diethanolamine	Triethanolamine	Reference
Molecular weight (g/mole)	61.09	105.14	149.19	a
Boiling point (°C at 760 nm Hg)	171	268	340	a
Freezing point (°C at 760 mm Hg)	10	28	21	a
Specific gravity (at 25/4 °C)	1.0113	1.0881 (at 30/4 °C)	1.1205	a
Viscosity (centipoise at 25 °C)	18.9	351.9 (at 30 °C)	600.7	a
Vapor pressure (pa)	1.00×10^2	6.31×10^{-1}	2.39×10^{-2}	Lyman et al. 1982
Henry's Law constant (atm·m^3/mol)	2.45×10^{-7}	5.35×10^{-14}	3.38×10^{-19}	Lyman et al. 1982
				Howard 1990
pH (10% aqueous solution at 25 °C)	12.00	11.45	10.8	Lewis 1992
pK_a	9.68	9.01	7.92	Lewis 1992
Solubilities				
Water (theoretical calculation; mg/L at 20 °C)	2.465×10^5	3.180×10^5	6.273×10^5	Lyman et al. 1982
Water (x g/100 g at 20 °C)	Completely miscible	Completely miscible	Completely miscible	a
log K_{ow}	−1.31	−1.43	−1.75	b
log BCF	−1.23	−1.32	−1.57	Lyman et al. 1982
K_{oc}	4.62	3.97	2.63	b
ThOD NH$_3$ (p O$_2$/p compound)	1.3096	1.5218	1.5656	Budavari 1989
ThOD HNO$_3$ (p O$_2$/p compound)	2.3573	2.1305	1.9975	Budavari 1989
ThOD NO (p O$_2$/p compound)	1.9644	1.9023	1.8356	Budavair 1989
Odor threshold (mg/L at 60 °C)	32	160	160	Alexander et al. 1982
Physical form (at 25 °C)	Liquid	Solid	Liquid	a

K_{ow}, octanol–water partition coefficient; BCF, bioconcentration factor; K_{oc}, adsorption coefficient; ThOD, theoretical oxygen demand; p, part.
[a]The Dow Chemical Company (1988) Physical properties of the alkanolamines. Form no. 111-1227-88. The Dow Chemical Company, Midland, MI.
[b]Medicinal Chemistry Project (1989) Med. Chem. release 3.54. Daylight Chemical Information Systems, Inc., Irvine, CA.

Table 3. Selected physical properties of the isopropanolamines.

Property	Isopropanolamine	Diisopropanolamine	Triisopropanolamine	Reference
Molecular weight (g/mole)	75.11	133.19	191.27	a
Boiling point (°C at 760 mm Hg)	159	249	306	a
Freezing point (°C at 760 mm Hg)	3	44	44	a
Specific gravity (at 20/4°C)	0.906	0.992 (at 40/4 °C)	0.988 (at 70/4 °C)	a
Viscosity (centipoise at 25 °C)	23	870 (at 30 °C)	34.57	a
Vapor pressure (pa)	1.90×10^{2}	2.33	1.61×10^{-1}	b
Henry's Law constant (atm·m^3/mol)	2.11×10^{-6}	1.72×10^{-7}	4.21×10^{-8}	b
pH (10% aqueous solution at 25 °C)	11.9	11.5	10.83	Lewis 1992
pK_a	9.66	9.10	8.06	Lewis 1992
Solubilities				
Water (theoretical calculation; mg/L at 20 °C)	6.709×10^{4}	1.780×10^{4}	7.205×10^{3}	Lyman et al. 1982
Water (x g/100 g at 20 °C)	Completely miscible	1200	>500	a
log K_{ow}	−0.960	−0.820	−0.819	b
log BCF	−0.76	−0.28	0.04	Lyman et al. 1982
K_{oc}	9.95	21.77	37.12	b
ThOD NH$_3$ (p O$_2$/p compound)	1.7041	1.9221	2.0077	Budavari 1989
ThOD HNO$_3$ (p O$_2$/p compound)	2.5562	2.4026	2.3423	Budavari 1989
ThOD NO (p O$_2$/p compound)	2.2367	2.2224	2.2168	Budavari 1989
Odor threshold (mg/L at 60 °C)	28	10	c	Alexander et al. 1982
Physical form (at 25 °C)	Liquid	Solid	Solid	a
Odor	Slight ammoniacal	c	c	Sax and Lewis 1987

[a]The Dow Chemical Company (1988) Physical properties of the alkanolamines. Form no. 111-1227-88. The Dow Chemical Company, Midland, MI.
[b]Medicinal Chemistry Project (1989) Med. Chem. release 3.54. Daylight Chemical Information Systems, Inc., Irvine, CA.
[c]Information not available.

Table 4. Selected physical properties of the butanolamines.

Property	Butanolamine (1-amino-2-butanol)	Dibutanolamine	Tributanolamine	Reference
Molecular weight (g/mole)	89.04	161.08	233.12	Long 1955
Boiling point (°C at 760 mm Hg)	164.57	256.16	310.0	Long 1955
Freezing point (°C at 760 mm Hg)	"Thick at 50 °C"	~12	~2	Long 1955
Specific gravity (at 25°C/4 °C)	0.9359	0.9653	0.9859	Long 1955
Viscosity (centipoise at 25 °C)	28.96	887.9	5883.5	Long 1955
Vapor pressure (pa)	1.56×10^2	1.79	1.18×10^{-1}	[a]
Henry's Law constant (atm·m^3/mol)	3.39×10^{-6}	2.93×10^{-7}	9.11×10^{-8}	[a]
pH (10% aqueous solution at 25 °C)	11.65	11.45	10.91	Long 1955
Solubilities				
Water (theoretical calculation; mg/L at 20 °C)	4.03×10^4	9.719×10^3	2.992×10^3	[a,b]
Water (x g/100 g at 25 °C)	Very soluble	Very soluble	13.40	Long 1955
log K_{ow}	−0.457	0.213	0.768	[a]
log BCF	−0.58	−0.07	0.35	Lyman et al. 1982
K_{oc}	13.44	31.11	62.34	[a]
ThOD NH$_3$ (p O$_2$/p compound)	1.9745	2.1831	2.2627	Budavari 1989
ThOD HNO$_3$ (p O$_2$/p compound)	2.6925	2.5800	2.5370	Budavari 1989
ThOD NO (p O$_2$/p compound)	2.4232	2.4311	2.4342	Budavari 1989

[a] Medicinal Chemistry Project (1989) Med. Chem. release 3.54. Daylight Chemical Information Systems, Inc., Irvine, CA.
[b] Information not available.

units), $ÆZ_b = 0.97$ (constant; no units), and $R = 1.987$ cal/mol °K. The calculated P_{vp} values for the alkanolamines ranged from a high of 1.56×10^2 (e.g., butanolamine) to a low of 2.39×10^{-2} (e.g., TEA). These vapor pressures demonstrate that alkanolamines have a relatively low rate of evaporation. The octanol–water coefficient (K_{ow}) for the alkanolamines was estimated based on a model developed at Pomona College (Medicinal Chemistry Project 1989) using the structure fragment method. The K_{ow} relates the concentration of a chemical in a two phase octanol–water system. This parameter has been shown to be important in relating a chemical's potential to both sorb to soils and sediments and bioconcentrate in aquatic organisms. The log K_{ow} for the alkanolamines had been estimated to be very low and in the range of -1.75 to 0.35. This range is typical for highly miscible materials with very low sorption potential (Gerstl 1990; Lyman et al. 1982).

Although the alkanolamines are reported to be completely miscible in water, discrete values for each compound are required by the fugacity model of Mackay and Paterson (1981). The aqueous solubilities of the alkanolamines were calculated using Eq. (2).

B. Water Solubility

$$\log S = (-0.922 \times \log K_{ow}) + 4.184, \tag{2}$$

where S is the solubility in mg/L. The calculated solubilities for the alkanolamines ranged from 7.20×10^3 mg/L (e.g., TIPA) to a high of 6.27×10^5 mg/L for TEA.

In general, the alkanolamines are completely miscible in water and have high water solubilities, low potential for volatilization from water, and relatively low K_{ow} values. Aqueous solutions of the alkanolamines are basic with the pK_as decreasing with increased alkyl substitution.

III. Environmental Distribution

A unit world model developed by Mackay and Paterson (1981) was used to examine partitioning of the alkanolamines into the various environmental compartments including air, water, soil, sediment, suspended aquatic matter, and biota. The model is based on the fugacity of a compound, which can be described as the "escaping tendency" of a compound from a particular phase. This model predicts the distribution of a compound at equilibrium in the various compartments based on temperature and the physical properties of the chemical. In this model, no attempt is made to simulate the actual environment. The aim of the model is to provide estimated behavioral information characteristic of the substance in a steady-state system. However, the rate of transport of the compound between various compartments, as well as the possibility of biotic or abiotic degradation of the compound, are not considered in this model.

A summary of the model predictions for environmental distribution of the alkanolamines is found in Table 5. This model assumes a fixed or constant concentration of the chemical added to the unit world with no further input or outflow from the system. Results from this model demonstrate that ethanolamines, isopropanolamines, and the butanolamines are likely to partition almost exclusively into the water compartment (Table 5). This tendency to remain in the aqueous phase increases with increasing substitution (e.g., mono-substituted < di-substituted < tri-substituted) of the amine group ranging from 89.2% to 99.1% for the mono-substituted compounds to >99% for the tri-substituted compounds. Based upon their low K_{ow}s, the potential for these materials to partition to soil or sediment is minimal (<0.1%) and would not represent a major environmental sink for the alkanolamines. However, the model simulations demonstrated limited partitioning ($\sim 5\%$–11%) of the mono-substituted compounds into the atmosphere.

This unit world model predicts that alkanolamines would partition primarily into the aqueous compartment at equilibrium, with the remainder distributed to the atmosphere. These predictions were based on the high aqueous solubilities, low K_{ow}s, and low P_{vp}s for the alkanolamines.

It is important to note that the unit world model predictions for partitioning of a chemical into the soil/sediment compartment are a function of the K_{ow} and water solubility, which is reasonable for most nonpolar organic species. However, for polar or ionizable compounds, chemical sorption to soil or sediment can involve other mechanisms. For example, recent studies with the ethyleneamines have shown that interaction of these protonated amines and negatively charged soil is possible (Davis 1993). Because the

Table 5. Environmental distribution of the alkanolamines predicted by the unit world model (Mackay and Paterson 1981).

Compound	Air (%)	Water (%)	Soil (%)	Sediment (%)	Suspended aquatic matter (%)	Biota (%)
Ethanolamine	0.86	99.14	0.00	0.00	0.00	0.00
Diethanolamine	0.01	99.99	0.00	0.00	0.00	0.00
Triethanolamine	0.00	100.00	0.00	0.00	0.00	0.00
Isopropanolamine	6.96	93.04	0.00	0.00	0.00	0.00
Diisopropanolamine	0.61	99.38	0.01	0.01	0.00	0.00
Triisopropanolamine	0.15	99.82	0.02	0.02	0.00	0.00
Butanolamine						
(1-amino-2-butanol)	10.81	89.18	0.00	0.00	0.00	0.00
Dibutanolamine	1.03	98.94	0.01	0.01	0.00	0.00
Tributanolamine	0.32	99.59	0.05	0.04	0.00	0.00

Total percentages may not equal 100% due to rounding of model.

alkanolamines would also be protonated at environmentally relevant pHs, they are likely to associate with soil to a greater extent than predicted by their water solubility and K_{ow} alone. Thus, the unit world model may likely underestimate the adsorption capacity of alkanolamines.

IV. Atmospheric Degradation
A. Ethanolamines

A very small percentage of the ethanolamines and isopropanolamines would be expected to partition into the atmosphere at equilibrium based upon the estimates from the Mackay and Paterson model. The residual concentrations of ethanolamines and isopropanolamines present in the atmosphere are expected to exist almost entirely in the vapor phase with some TEA sorbed to atmospheric particles. The dominant removal mechanism is expected to be reaction of these compounds with photochemically generated hydroxyl radicals (Atkinson 1988; Muller and Klein 1991). The high solubility of the ethanolamines and isopropanolamines suggests that they may be removed from the atmosphere in precipitation. Dry deposition also may be an important removal process for triethanolamine (Eisenreich et al. 1981). The half-life of the ethanolamines and monoisopropanolamine (MIPA) was estimated using the Syracuse Research Corp. Atmospheric Oxidation Program Version 1.3. (Howard 1991). On the basis of this program, the half-life of these compounds in the atmosphere was estimated to be approximately 4 hr for the ethanolamines and approximately 10 hr for MIPA.

V. Bioconcentration Potential

Certain chemicals, especially the more hydrophobic compounds, tend to partition from the water column and accumulate in aquatic organisms. This accumulation, or bioconcentration, is of concern because it can lead to elevated concentrations of a chemical in the food chain. The propensity for a compound to bioaccumulate is referred to as the bioconcentration factor (BCF), defined as the concentration of the chemical in the organism at equilibrium divided by the concentration of the chemical in water.

Since bioaccumulation is an important property, there is a need for screening materials and chemicals for their potential to accumulate in the food chain. BCF values are routinely estimated from quantitative structure–activity relationships, although experimental values for BCF are preferred (Lyman et al. 1982). The accumulation of a chemical occurs primarily in the fatty tissue of aquatic organisms and has been shown to be a function of the lipid solubility of a material. Due to this correlation between lipid solubility and bioconcentration, the K_{ow} for a chemical has been used to estimate BCF. The relationship between K_{ow} and BCF has been described by Eq. (3) (Lyman et al. 1982). Values for log BCF reported in this review were determined based on Eq. (3):

$$\log \text{BCF} = (0.76 \times \log K_{ow}) - 0.23. \tag{3}$$

Log BCF values for the ethanolamines, isopropanolamines, and butanolamines are presented in Table 6. These BCF values range from -1.56 to 0.35, with most estimated to be less than 0. BCF values relate the magnitude of uptake of a chemical by aquatic organisms. In general, the potential for bioconcentration of the alkanolamines is extremely low since the BCF values were routinely less than 5. These low values are not surprising since the estimated K_{ow} values for the same materials were also determined to be very low. The low K_{ow} and BCF values reported for the alkanolamines indicate that these compounds are not likely to accumulate in the aquatic food chain (Thomann 1989).

VI. Reactions of the Alkanolamines

The alkanolamines are bifunctional molecules having both amino and alcohol functional groups. The amino group may be either primary, secondary, or tertiary. The alcohol functional grouping may be either primary, as in the case of the ethanolamines, or secondary, as in the isopropanolamines and butanolamines.

Because they are bifunctional, the alkanolamines undergo many types of reactions. Alkanolamines can react with carbon dioxide (CO_2) and hydrogen sulfide in aqueous solution to form salts, a reaction used in the "sweetening" of natural gas. They can also form neutral alkanolamine soaps when combined with long-chain fatty acids. These compounds are often used as

Table 6. Bioconcentration factors (BCF) for ethanolamines, isopropanolamines, and butanolamines.

Compound	log BCF
Ethanolamines	
Monoethanolamine	-1.23
Diethanolamine	-1.32
Triethanolamine	-1.56
Isopropanolamines	
Isopropanolamine	-0.76
Diisopropranolamine	-0.28
Triisopropanolamine	0.04
Butanolamines	
Butanolamine	-0.58
Dibutanolamine	-0.07
Tributanolamine	0.35

emulsifiers. Alkanolamines also react with aldehydes, ketones, alkyl and aryl halides, and epoxides to form other industrially significant compounds. The alkanolamines have also been reported to react with nitrosating compounds such as NO_2, N_2O_3, or N_2O_4 to form nitrosamines. In most cases, nitrosamines are formed under acidic conditions; however, nitrosation reactions have been reported to occur up to pH 11 (Anderson 1979).

Although there is abundant information available on the toxicology of nitrosamines and their formation in animals (Mirvish 1975), there are conflicting reports on the potential for the generation of these compounds in natural ecosystems. There have been limited reports that biological processes may play a role in the formation of nitrosamines in the environment. Nitrosamine formation has been reported to be mediated by microorganisms, either enzymatically, by other cell constituents, or by organic matter present in soils and sewage sludge (Ayanaba and Alexander 1974; Mills and Alexander 1976; Richardson et al. 1979; Yordy and Alexander 1981). Yordy and Alexander (1981) reported the transient formation of low levels of N-nitrosodiethanolamine (N-DELA) from diethanolamine in lake water and raw sewage. They suggested a role for microorganisms (or a heat-labile agent) in the generation of N-DELA. However, in this study only a limited number of samples were examined, and N-DELA was formed only on a transient basis and was shown to undergo further transformation (e.g., biodegradation) in both lake water and sewage (Yordy and Alexander 1980).

Demonstration of the potential for biologically mediated N-nitrosamine formation in the environment was not reproducible in subsequent studies using similar amine compounds (Bailey et al. 1991; Gonsior and West 1991). Bailey et al. (1991) examined the potential for N-nitrosamine formation from the biodegradation of diethylenetriamine (DETA) in a variety of environmental matrices including sewage, lake water, and soil. This study was required by the U.S. Environmental Protection Agency (EPA) under the Toxic Substances Control Act and focused on screening for all possible N-nitrosamines that could be formed from DETA. Specifically, Bailey et al. focused on developing analytical techniques that would enable detection of all possible N-nitrosamines that could be formed from DETA. This analytical method resulted in separation of DETA from its impurities and potential metabolites, allowing detection of N-nitrosamine at levels of ≥ 500 μg/L. N-nitrosamine analyses of both water and activated sludge samples were conducted over the entire course of the study. These results clearly demonstrated that N-nitrosamines were not formed during the biodegradation of DETA in sewage, lake water, or soil at levels ≥ 500 μg/L.

Two additional studies were conducted by Gonsior and West (1991) and West (1995) to examine the potential for N-nitrosamine formation during the biodegradation of triethanolamine and diisopropanolamine in activated sludge. Samples of activated sludge reactions containing either TEA or DIPA were analyzed for the presence of N-nitrosamine by differential pulse polarography. The addition of TEA or DIPA to activated sludge did not result in the formation of *N*-nitrosamines above a detection limit of 100 $\mu g/$ L.

Studies using ^{14}C-radiolabeled TEA (Gonsior and West 1991) and DIPA (West 1995) showed no evidence for the creation of N-nitrosamines during biodegradation of TEA or DIPA in soil and river water. Analytical methods used in these studies were not specific for *N*-nitrosamines but did allow detection of any radiolabeled metabolic degradation products of TEA or DIPA at levels $\geq 2\%$ of the initial parent material. Although TEA and DIPA were shown to biodegrade in both soil and lake water, intermediate degradation products were transitory and did not persist at levels $\geq 2\%$ of the initial applied material.

VII. Biodegradation

Because the alkanolamines are expected to partition primarily into the water column, understanding their expected fate and lifetime in aquatic environments is very important. Organic chemicals can be removed from the water column in a variety of ways, including adsorption, volatilization, physical–chemical interaction, and biodegradation. Biodegradation has been shown to be one of the dominant processes impacting the fate of a chemical in the environment (Alexander 1985; Klecka 1985). Biodegradation is a key process for limiting and reducing the exposure concentration of a chemical to the biota.

A wide range of laboratory tests have been developed to examine biodegradation of organic chemicals (Howard and Banerjee 1984; OECD 1993). The primary objective of these tests is to assess the extent to which a microorganism can transform a chemical. The underlying assumption of the biodegradation tests is that the results from these laboratory studies will provide relevant information on the environmental fate of a chemical. Information derived from these tests has been used to predict degradation rates and estimates for the reduction in total mass of a chemical from specific environmental compartments. However, biodegradation testing is not an end in itself but should be included as part of an overall environmental safety assessment.

The Organization for Economic Cooperation and Development (OECD) has classified biodegradation testing methods into a sequential or tiered system, beginning with relatively inexpensive, fast, screening tests and progressing toward more complex testing methods (OECD 1993). This tiered approach assists in identifying those compounds that may be recalcitrant to biodegradation and have extended lifetimes in the environment. A partial list of standardized biodegradation tests is provided in Table 7. Biodegradation tests can be grouped into three major categories: screening tests, tests that measure inherent biodegradation, and simulations tests. In general, these tests differ in terms of the source, concentration, and acclimation

Table 7. Biodegradation testing methods.

Test	Represents	Duration (d)	Inoculum and concentration	Concentration (mg C/L)	Analysis
A. Tests for ready biodegradability					
Closed-bottle test	Surface water	30	≤5 mL of effluent/L	2–10	DO
Biochemical oxygen demand test (BOD)	Surface water or sewage treatment	5–28	Varied (low)	3–20	DO
Modified OECD screening test	Surface water	28	0.05% of either sewage effluent, surface water, or soil extract	10–40	DOC
Modified Sturm Test	Surface water	28	1% municipal sewage effluent	10–20	DOC and CO_2
French AFNOR test	Polluted river water	42	$(5 \pm 3) \times 10^5$ bacteria/mL	40	DOC
Japanese MITI test	Sewage treatment plant	28	30 mg sludge/L	50–100	DOC and DO
B. Tests for inherent biodegradability					
Zahn–Wellens	Sewage treatment plant	28	Activated sludge dry solids (0.2–1.0 g/L)	~400	DOC
Modified SCAS	Sewage treatment plant		Activated sludge dry solids (2.5 g/L)	5–20	DOC
C. Simulation tests					
Coupled units	Continuous completely mixed activated sludge process	Varied	Activated sludge dry solids (2.5 g/L)	12	DOC

DO, Dissolved oxygen; OECD, Organization for Economic Cooperation and Development; DOC, dissolved organic carbon; AFNOR, Association Francaise de Normalisation; MITI, Ministry for International Trade and Industry; SCAS, Semi-continuous activated sludge.

of the inoculum, the concentration of the test compound, and analytical measurement techniques used.

The first level of biodegradation testing includes the screening test for "ready biodegradability." These tests are easy to perform, of short duration, and relatively inexpensive. A large number of compounds can be routinely screened using this methodology. In a screening test, biodegradation is examined under stringent conditions, often involving low concentrations of the test chemical serving as the sole carbon and energy source and an unadapted or nonacclimated inoculum. Results from these tests are usually presented as percentage of biodegradation after a specified time interval and are expressed as percentage of dissolved organic carbon (DOC) removal, percentage of CO_2 production, or oxygen consumed as a percentage of the theoretical oxygen demand (ThOD). A compound is defined as "readily biodegradable" by the OECD guidelines if 70% DOC removal, 60% removal of theoretical oxygen demand (ThOD), or 60% production of theoretical CO_2 is achieved within a 10-d period after the start of biodegradation (OECD 1993).

If a ready test is passed, the compound is not expected to persist in the environment. However, failure to pass a screening test does not indicate that the material is resistant to biodegradation but may indicate that an acclimation period is required or that the material was toxic to the inoculum. A compound failing the screening tests would then be tested for inherent biodegradation. A positive result in an inherent test demonstrates that a chemical has the potential to undergo biodegradation given the right circumstances.

Inherent and screening tests can also be used to evaluate primary and ultimate degradation. Primary biodegradation occurs when the parent material is transformed such that the basic properties of the chemical are lost but may not include complete conversion to inorganic materials (e.g., CO_2 and salts). Complete conversion of a chemical to biomass and inorganic end products such as CO_2 and water is referred to as *mineralization* or *ultimate biodegradation*.

Biodegradation of the ethanolamines has been extensively studied in a variety of screening tests designed to measure ready biodegradability. The most comprehensive set of data available for these compounds are the results from the biochemical oxygen demand test (BOD). Historically, the BOD test has been used to provide information on waste loading and treatment efficiency for waste treatment facilities (Gaudy 1972). BOD relates the amount of oxygen consumed by microorganisms during degradation of organic matter as a relative indication of biodegradability. BOD values are routinely reported on the basis of the incubation period of the test, e.g., 5-d BOD (BOD_5). Several independent laboratories have examined the biodegradation of alkanolamines using the BOD test and have reported a wide range of BOD values (Table 8). MEA has reported BOD values of 40%–67% (Alexander and Batchelder 1975; Bridie et al. 1979a; Lamb and Jenkins 1952; Mills and Stack 1952; Urano and Kato 1986; Young et al. 1968) while BOD values for TEA and DEA ranged from a low of approximately 7% to a high of 88% (Gerike and Fisher 1979; Price et al. 1974). The range of BOD values observed for the ethanolamines was most likely due to the fact that the BOD tests were conducted by several independent laboratories using microorganisms from widely different sources. The variability inherent in BOD testing, as well as many other screening tests for biodegradation, has been attributed to the inoculum (Gerike and Fisher 1979; Struijis et al. 1995). Although BOD values for these compounds spanned a wide range, they were routinely greater than 60%. These high values demonstrate that, even under the stringent conditions of the BOD test, the ethanolamines are biodegradable.

Ethanolamines were also shown to undergo biodegradation in screening tests that generally follow the OECD guidelines for ready biodegradability. Routinely, MEA, DEA, and TEA have been reported to undergo extensive biodegradation using the OECD test methods. Degradation of the ethanolamines was usually complete (e.g., mineralization) since the DOC loss from the test system was typically $> 90\%$ (see Table 8), although CO_2 production was not always monitored (Birch and Fletcher 1991; Huntziger et al. 1978). In one study in which CO_2 production was reported, 92% of MEA degradation could be accounted for as CO_2 (Kuenemann et al. 1992).

A series of inherent biodegradation tests have been conducted by several laboratories to evaluate the biodegradation potential of ethanolamines (see Table 8). Test methods for inherent biodegradation are less stringent than the screening tests and generally include conditions more favorable to biodegradation. Often the inherent test includes a higher concentration of inoculum, which provides a greater diversity of microorganisms. Inherent biodegradation tests can include an acclimation or preconditioning of the inoculum, and this acclimation has been reported to increase the likelihood for biodegradation (Sugatt et al. 1984). As a result, both the extent and rate of biodegradation of ethanolamines was greater in inherent tests than reported for the screening studies (Table 8). For example, the BOD_5 for DEA was $\leq 30\%$ using an unacclimated inoculum. However, the BOD_5 for DEA increased to 77%–97% when an acclimated inoculum was included in the test (Bridie et al. 1979a; Gannon et al. 1978). In a separate series of laboratory studies using the OECD inherent tests, degradation of MEA, DEA, and TEA approached 100% using an acclimated inoculum (Gerike and Fisher 1979). The test materials were almost completely mineralized to CO_2 with levels reported consistently $> 90\%$.

There have been only a few reports from screening or inherent studies examining the biodegradation of the isopropanolamines (MIPA, DIPA, and TIPA). In the BOD test, some biodegradation was noted for all three isopropanolamines, with BOD values reported to range from $\sim 40\%$ to 50% (Batchelder and Rhinehart 1977; Bridie et al. 1979a). As with the ethanolamines, acclimation of the inoculum appeared to result in an in-

Table 8. Biodegradability of the alkanolamines.
A. Tests for Ready Biodegradability

BOD Tests	Percent degradation[a] (d)[b]				Comments	Reference
	5 d	10 d	20 d	>20 d		
Monoethanolamine	34		40		Sewage inoculum Degradation occurred after a lag period of 2 d	Young et al. 1968
	0	58.4	64	75	Filtered sewage seed 2.5 ppm initial concentration	Lamb and Jenkins 1952
	71				Filtered sewage seed inoculum	Bridie et al. 1979a
	40				Sewage inoculum 100 ppm initial concentration 14% DOC removal Electrolytic respirometer	Urano and Kata 1986
		65			Sewage inoculum	Mills and Stack 1952
	38	48	67			Alexander and Batchelder 1975
Diethanolamine	0.9	1.4	6.8		Sewage inoculum 2.5 ppm initial concentration	Lamb and Jenkins 1952
	17	72	88		Sewage inoculum	Price et al. 1974
	2	60	76		Sewage inoculum Synthetic seawater matrix	Price et al. 1974
	2				Effluent from "biological waste treatment plant" as inoculum	Bridie et al. 1979a

				Conditions	Reference
	0			Sewage inoculum 100–1000 ppm initial concentration	Mills and Stack 1952
			10	Sewage inoculum 2 ppm initial concentration	Gerike and Fisher 1979
			94	Enriched sewage inoculum 2 ppm initial concentration	Gerike and Fisher 1979
	11	36	58		Alexander and Batchelder 1975
Triethanolamine	0	0.8	6.2	Sewage inoculum 2.5 ppm initial concentration	Lamb and Jenkins 1952
	8	9	66	Sewage inoculum	Price et al. 1974
	56	65	69	Sewage inoculum Synthetic seawater matrix	Price et al. 1974
	5	0		Sewage inoculum	Mills and Stack 1952
				Effluent from biological sanitary waste treatment plant used as inoculum	Bridie et al. 1979a
			0	Sewage inoculum 2 ppm initial concentration	Gerike and Fisher 1979
			9	Enriched sewage inoculum 2 ppm initial concentration	Gerike and Fisher 1979
	0	0	61		Alexander and Batchelder 1975
Isopropanolamine	4	34.0	46.0	Sewage inoculum	Bridie et al. 1979a
	5.1	37	38	Sewage inoculum	Lamb and Jenkins 1952
					Alexander and Batchelder 1975

(continued)

Table 8. (*Continued*)

BOD Tests	Percent degradation[a] (d)[b]				Comments	Reference
	5 d	10 d	20 d	>20 d		
Diisopropanolamine	0	0	39			Alexander and Batchelder 1975
Triisopropanolamine	0	46	0 >46			Alexander and Batchelder 1975 Batchelder and Rhinehart 1977

Modified OECD Screening Test	Percent DOC removal	Comments	Reference
Monoethanolamine	94 (28 d)	Sewage inoculum Initial concentration ~20 mg/L DOC	Kuenemann et al. 1992
Diethanolamine	100 (19 d)	Sewage inoculum 3–20 ppm DOC initial concentration	Gerike and Fisher 1979
Triethanolamine	96 (19 d)	Sewage inoculum 3–20 ppm DOC initial concentration	Gerike and Fisher 1979

Modified Sturm Test	Percent DOC removal after 28 d	Percent mineralization to CO_2 after 28 d	Comments	Reference
Monoethanolamine	97	92	Sewage inoculum	Kuenemann et al. 1992
	—	91.8	Secondary effluent as inoculum	Birch and Fletcher 1991

Modified AFNOR Test	Percent DOC removal		Comments	Reference
	28 d	42 d		
Monoethanolamine			No information available	
Diethanolamine	97	98	Sewage inoculum 40 ppm DOC initial concentration	Gerike and Fisher 1979
Triethanolamine	—	97	Sewage inoculum 40 ppm DOC initial concentration	Gerike and Fisher 1979

(continued)

Table 8. (*Continued*)

MITI Test	Percent of ThOD[c] (14 d)	Percent DOC removal (14 d)	Comments	Reference
Monoethanolamine		Passed		Huntziger et al. 1978
Diethanolamine	3	0	Activated sludge inoculum 50 ppm initial concentration	Gerike and Fisher 1979
Triethanolamine		Passed		Huntziger et al. 1978
	2	0	Activated sludge inoculum 100 ppm initial concentration	Gerike and Fisher 1979

B. Tests for Inherent Biodegradability

BOD Tests[d]	Percent degradation[a] (d)[b]				Comments	Reference
	5 d	10 d	20 d	>20 d		
Monoethanolamine					No information available	
Diethanolamine	97				Bacteria isolated from cutting fluid and a sewage population. Various concentrations. Maximum oxidation at 500 ppm. Inhibition at 2000 mg/L	Gannon et al. 1978

		Comments	Reference
Triethanolamine	77	Effluent from biological waste treatment plant used as inoculum (acclimated)	Bridie et al. 1979a
	28	Effluent from biological waste treatment plant used as inoculum (acclimated)	Bridie et al. 1979
	22	Activated sludge inoculum 500 ppm initial concentration 15 d acclimation	Gannon et al. 1978
Isopropanolamine Diisopropanolamine	43	Acclimated sewage inoculum	Bridie et al. 1979a
Triisopropanolamine	51 >75	No information available	Batchelder and Rhinehart 1977

	Percent DOC removal (d)	Comments	Reference
Modified OECD Screening Test[a]			
Monoethanolamine	99 (28 d)	Preconditioned sewage inoculum Initial concentration ~20 mg/L DOC	Kuenemann et al. 1992

(continued)

Table 8. (*Continued*)

Modified Sturm Test[d]	Percent DOC removal after 28 d	Percent mineralization to CO_2 after 28 d	Comments	Reference
Monoethanolamine	96	62	Preconditioned sewage inoculum Initial concentration ~20 mg/L DOC	Kuenemann et al. 1992
Diethanolamine	97	88	Activated sludge inoculum Initial concentration 10 ppm DOC 14 d acclimation	Gerike and Fisher 1979
Triethanolamine	100	91	Activated sludge inoculum Initial concentration 10 ppm DOC 14 d acclimation	Gerike and Fisher 1979

Zahn-Wellens Test	Percent DOC removal after 14 d	Comments	Reference
Monoethanolamine		No information available	
Diethanolamine	94	Initial concentration 400 ppm Activated sludge inoculum	Gerike and Fisher 1979

Triethanolamine		89	Initial concentration 1000 ppm COD Acclimated sewage inoculum 3-d lag period	Zahn and Wellens 1980
		82	Initial concentration 400 ppm Activated sludge inoculum	Gerike and Fisher 1979
Isopropanolamine		44 (24 d)	Initial concentration 500 ppm Activated sludge inoculum 10-d lag period	Zahn and Wellens 1980

C. Simulation Tests

Compound	Test method	Initial compound concentration	Inoculum	Rates/comments	Reference
Monoethanolamine				No information available	
Diethanolamine	Die-away	21 ppm	River water and *Pseudomonas* sp.	5% mineralization to CO_2 at 4 d	Boethling and Alexander 1979
		210 ppb	River water and *Pseudomonas* sp.	55% mineralization to CO_2 at 4 d	Boethling and Alexander 1979
		21 ppt	River water and *Pseudomonas* sp.	32% mineralization at CO_2 at 4 d	Boethling and Alexander 1979
	Coupled units	12 ppm DOC	2500 ppm MLSS	94% DOC removal in 7 d	Gerike and Fisher 1979

(continued)

Table 8. (*Continued*)

Compound	Test method	Initial compound concentration	Inoculum	Rates/comments	Reference
	Die away	1ppm	Lake water	31% mineralization to CO_2 at 14 d	Yordy and Alexander 1981
		1 ppm	Lake water	1.2% mineralization to CO_2 at 14 d (acidic lake)	Yordy and Alexander 1981
	Die-away	1ppm	Sewage inoculum	53% mineralization to CO_2 at 20 d	Yordy and Alexander 1981
	Die-away	50 ppm	River water and sewage	10-d acclimation period Half-life 7 d 90% BOD	Mills and Stack 1954
	Batch test		Acclimated activated sludge	Rate: 19.5 mg COD/g·hr 97% removal	Pitter 1976
Triethanolamine	Die-away	50 ppm	River water and sewage	28-d acclimation period 70% BOD of ThOD at 10 d	Mills and Stack 1954
	Coupled units	12 ppm DOC	2500 ppm $MLSS^e$	91% DOC removal in 7 d	Gerike and Fisher 1979
	Batch system	0.7 ppm	Soil and mineral salts	$t_{1/2} = 1.7$ d Mineralization rate: 0.406 d^{-1}	West and Gonsior 1996
		100 ppm	Soil and mineral salts	$t_{1/2} = 1.4$ d Mineralization rate: 0.498 d^{-1}	
		1000 ppm	Soil and mineral salts	$t_{1/2} = 5.4$ d Mineralization rate: 0.129 d^{-1}	
	Batch system	0.6 ppm	Municipal activated sludge (164 mg/L MLSS)	$t_{1/2} = 9$ hr Mineralization rate: 1.92 d^{-1}	

(continued)

Compound	Test	Concentration	Medium	Result	Reference
		5.7 ppm	Municipal activated sludge (164 mg/L MLSS)	$t_{1/2}$ = 16 hr, Mineralization rate: 1.04 d^{-1}	
		0.6 ppm	Municipal activated sludge (818 mg/L MLSS)	$t_{1/2}$ = 6 hr, Mineralization rate: 2.86 d^{-1}	
		5.7 ppm	Municipal activated sludge (818 mg/L MLSS)	$t_{1/2}$ = 10 hr, Mineralization rate: 1.69 d^{-1}	
	Batch system	100 ppb	River water	$t_{1/2}$ = 1.9 d, Mineralization rate: 0.359 d^{-1}, Lag period of 10–14 d	
		500 ppb	River water	$t_{1/2}$ = 1.2 d, Mineralization rate: 0.558 d^{-1}, Lag period of 6–10 d	
		100 ppb	River water + sediment	$t_{1/2}$ = 1.1 d, Mineralization rate: 0.385–0.724 d^{-1}	
		500 ppb	River water + sediment	$t_{1/2}$ = 1.8–7.2 d, Mineralization rate: 0.096–0.563 d^{-1}	
Isopropanolamine	Batch	~160 ppm	Anaerobic digester sludge	22 mg L^{-1} d^{-1}	Chou et al. 1979
Diisopropanolamine	Batch system	10 ppm–1000 ppm	Soil	$t_{1/2}$ = 10 to 27 d, 68–76% mineralization to CO_2	West 1995
	Batch system	0.5 ppm–5 ppm	Municipal activated sludge	$t_{1/2}$ = 1.1 to 2.3 d, 57–86% mineralization to CO_2	

Table 8. (*Continued*)

Compound	Test method	Initial compound concentration	Inoculum	Rates/comments	Reference
	Batch system	0.6 ppm	River water	$t_{1/2} = 2.7$ d 57–86% mineralization to CO_2	Cleveland and Ulmer 1995
Triisopropanolamine	Batch system	3.3 ppm	Soil	$t_{1/2} = 2$ d 66–72% mineralization to CO_2	
	Batch system	2.3	Lake water/sediment	$t_{1/2} = 14.3$ d 62% mineralization to CO_2	Krieger 1995

OECD, Organization for Economic Cooperation and Development; AFNOR, Association Francaise de Normalisation; MITI, Ministry for International Trade and Industry.

[a]Biochemical oxygen demand (BOD) is defined as parts of oxygen consumed per part of compound during degradation. This value is expressed as a percentage of the total (TOD) or theoretical (ThOD) oxygen demand. DOC, Dissolved oxygen concentration.

[b]Data reflect percent of degradation at or before the day in the column heading.

[c]ThOD refers to the theoretical oxygen demand of the compound and is calculated on its complete mineralization to inorganic products.

[d]This test is normally classified as a screening test. However, because the inoculum in these specific determinations was either preconditioned to the test system or acclimated to the test compound prior to the start of the test period, they are considered tests for inherent biodegradability.

[e]MLSS, Mixed liquor suspended solids.

creased rate and extent of isopropanolamine degradation. For example, the BOD_5 for MIPA and TIPA was usually $<5\%$ using an unacclimated inoculum. However, BOD_5 values for these compounds increased to 40%–50% when an acclimated inoculum was used in the same test.

There is little of information available on the anaerobic biodegradation of the isopropanolamines. However, MIPA was reported to undergo degradation by anaerobic sludge in a study simulating an anaerobic wastewater system. Approximately 65% of MIPA was degraded at a rate of 22 mg L^{-1} d^{-1}, although a lag phase of 9 d preceded degradation (Chou et al. 1979).

The screening and inherent biodegradation tests demonstrate that ethanolamines and isopropanolamines are biodegradable and would not be expected to persist in most aerobic environmental compartments. Because these tests are often used in determining the environmental acceptability of a material, it is important to understand the types of information provided by these methods. Screening and inherent tests can be used to provide conservative estimates of biodegradation but are usually not sufficient to provide meaningful kinetic data. Quantitative application of the information from these tests across different environmental compartments to predict actual environmental fate should be done with caution.

A more realistic approach for assessing environmental fate of a chemical is through the use of simulation tests. These laboratory tests are performed with actual environmental samples using realistic concentrations of the test chemical. These studies often use radiolabeled (e.g., ^{14}C) test chemicals to assist in determining mass balances and can provide unequivocal evidence of biodegradation. By more closely simulating actual environmental conditions, they provide information on rate and extent of biodegradation in specific environmental compartments. These environments can include surface water and groundwater, soil and river sediment, or wastewater treatment facilities. Specific simulation tests to assess environmental fate should be designed based on the expected use pattern of the chemical and should simulate the particular compartment of the receiving environment. That is, selection of the appropriate test methods should be related to the probable environmental distribution of the chemical.

Simulation studies to examine the fate and lifetime of selected alkanolamines have been conducted in activated sludge, soil, and freshwater aquatic systems (Table 8). Both the rate and extent of primary biodegradation and mineralization were routinely determined in multiple environmental matrices. Biodegradation studies using microorganisms from either municipal or industrial waste treatment facilities were used to evaluate biodegradation of ethanolamine and isopropanolamines. Results from biodegradation tests involving activated sludge or other wastewater treatment simulations indicated that the alkanolamines, in general, would undergo extensive transformation and removal in most waste treatment systems. For example, results from the coupled units test using activated sludge at ~2500 mg/L demonstrated >90% removal of TEA and DEA, as DOC, after 7 d

(Gerike and Fisher 1979; Pitter 1976). Using ^{14}C-TEA, West and Gonsior (1996) observed rapid mineralization of TEA to $^{14}CO_2$ with half-lives in the range of 6–16 hr. Degradation rates observed in this study were reported to be directly correlated with the biomass levels (e.g., activated sludge). West and Gonsior used dilute activated sludge, and biodegradation of TEA is expected to occur at a greater rate in actual waste treatment plants due to the higher sludge levels typically present in these facilities.

The fate of diisopropanolamine has been examined in simulation tests using activated sludge. In dilute activated sludge (< 900 mg/L solids), West (1995) observed rapid degradation of ^{14}C-DIPA to CO_2 with complete removal of DIPA in 72–120 hr. Since DIPA has been shown to be a major metabolite during the aerobic biodegradation of TIPA (Cleveland and Ulmer 1995), one would also expect TIPA to behave in similar fashion and undergo complete mineralization in activated sludge systems. Note that the concentration of the alkanolamines in these studies was routinely < 12 ppm and reflects the levels that would be observed in most waste treatment facilities.

The impact of higher concentrations of ethanolamines on activated sludge was examined using the OECD inhibition test. This test is used to evaluate the potential impact of a chemical on wastewater treatment systems. These tests demonstrated that the IC_{50} for MEA, DEA, and TEA were all at a level greater than 1000 mg/L (Table 9) (Klecka and Landi 1985). The IC_{50} is defined as the concentration of the test chemical that causes a 50% inhibition of the respiration rate of activated sludge as compared to a control. Taken together, these data indicate that ethanolamines and isopropanolamines could be easily treated in typical wastewater treatment plants and should not prove inhibitory to the activated sludge microorganisms.

A series of related studies have been completed examining the fate and lifetime of selected ethanolamines and isopropanolamines in aerobic soil and aquatic environments (Boethling and Alexander 1979; West 1995; Yordy and Alexander 1981). In freshwater systems, low levels of ethanolamines and isopropanolamines were reported to be mineralized to CO_2. The amount of alkanolamine mineralized ranged from 1% to 70% (see Table

Table 9. Activated sludge inhibition of the alkanolamines.

OECD Activated Sludge Inhibition Test	3-hr IC_{50}[a] (mg/L)	Reference
Monoethanolamine	> 1000	Klecka and Landi 1985
Diethanolamine	> 1000	Klecka and Landi 1985
Triethanolamine	> 1000	Klecka and Landi 1985

[a]Activated sludge inhibition test measures potential impact of a chemical on wastewater treatment systems. The IC_{50} is the concentration of the test chemical that causes a 50% inhibition of respiration rate of activated sludge compared to control.

8). This variability in mineralization can likely be attributed to the source of the lake or river water inoculum. In some of theses studies (Yordy and Alexander 1981), the mineralization rates may have been limited by either low biomass levels or adverse environmental conditions (e.g., low pH). In spite of this variability, the degradation half-lives for these materials were comparable in either lake or river water environments, which indicates that the organisms responsible for degradation of the alkanolamines are widespread in the aquatic environment. For example, the half-life for TEA in two limnologically distinct Michigan river waters was 1.6 d (West and Gonsior 1996). In a separate study, the biodegradation of TIPA in a freshwater pond sediment system was somewhat slower, with a predicted half-life of 14.3 d (Krieger 1995). Again, the use of ^{14}C-TIPA provided definitive evidence that degradation was complete with recovery of $^{14}CO_2$ around 64% (Table 8).

The fate of ethanolamines and isopropanolamines in soil is of interest because these compounds have been used in the formulation of agricultural chemicals such as pesticides. Several environmental fate studies have been conducted using ^{14}C-TEA, -DIPA, and -TIPA in a variety of soil environments (Cleveland and Ulmer 1995; West 1995; West and Gonsior 1996). Mineralization of the ethanolamines and isopropanolamines was observed in soil with recovery of $^{14}CO_2$ routinely greater than 50%. Half-lives for TEA and TIPA in soil were in the range of 1–6 d. The half-life for DIPA was reported to be somewhat longer, 10–27 d, although a 1- to 2-wk lag period may account for the longer time required for biodegradation to occur in this study (West 1995).

In summary, a variety of ready, inherent, and simulation tests have indicated that alkanolamines are readily susceptible to biodegradation, with half-lives routinely in the range of 1 d to 2 wk. Biodegradation of these compounds is often complete, with CO_2 as the dominant degradation product. Both the rate of degradation and the extent of mineralization have been reported to increase with acclimation. A series of simulation tests have shown that the microorganisms responsible for degradation of the alkanolamines are fairly ubiquitous in activated sludge, soil, and freshwater aquatic environments. Based on these studies, one would not expect the alkanolamines to persist in most aquatic or terrestrial aerobic environments.

The use of pure cultures of microorganisms has been a valuable tool in elucidating the metabolic pathway of degradation of a wide range of organic chemicals. A variety of diverse microorganisms have been isolated that are capable of degrading the alkanolamines. Several pure cultures have been identified that utilize ethanolamine as the sole nitrogen, energy, and carbon source (Jones and Turner 1973; Jones et al. 1973).

Several metabolic pathways have been suggested by which microorganisms may degrade ethanolamines. One proposed pathway, associated with a gram-negative, rod-shaped bacterium, is detailed in Fig. 1 (Williams and

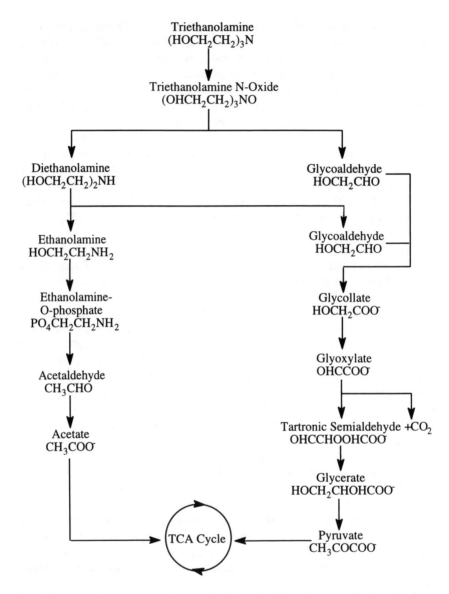

Fig. 1. Proposed pathway for the metabolism of triethanolamine, diethanolamine, and ethanolamine by a gram-negative rod (Williams and Calley 1981). *TCA*, tricarboxylic acid.

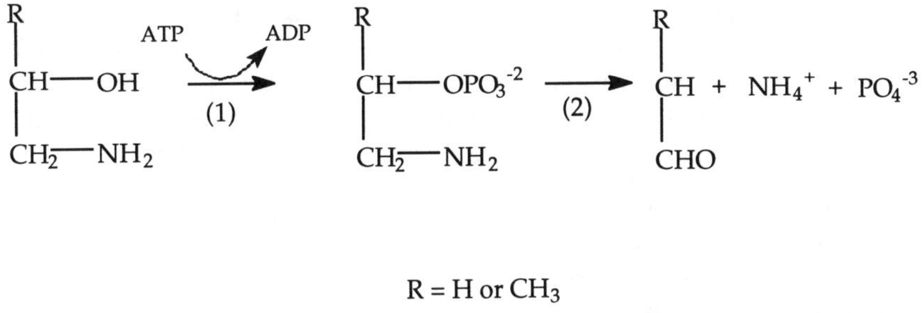

$$
\begin{array}{ccc}
\underset{\substack{| \\ CH_2{-}NH_2}}{\overset{\substack{R \\ |}}{CH{-}OH}} & \xrightarrow[\text{(1)}]{\text{ATP} \quad \text{ADP}} & \underset{\substack{| \\ CH_2{-}NH_2}}{\overset{\substack{R \\ |}}{CH{-}OPO_3^{-2}}} \xrightarrow[\text{(2)}]{} \underset{\substack{| \\ CHO}}{\overset{\substack{R \\ |}}{CH}} + NH_4^+ + PO_4^{-3}
\end{array}
$$

$$R = H \text{ or } CH_3$$

Fig. 2. Proposed pathway for the metabolism of amino alcohols by *Erwinia caroto-vora* (Jones and Turner 1973). Key to enzymes: (*1*) ATP-amino alcohol phosphotransferase (kinase); (*2*) amino alcohol *O*-phosphate phosphorylase (deaminating).

Calley 1981). This microorganism was isolated from activated sludge and was able to utilize the ethanolamines as the sole source of carbon for growth and energy. The putative degradative enzymes appear to be inducible since bacterial cultures exposed to the ethanolamines contained increased levels of the corresponding degradative enzymes compared to those cultures not exposed. This degradative pathway as outlined is very similar to the pathway reported to be present in the yeast *Rhodoturula mucilaginosa* (Fattakhova et al. 1991) and an anaerobic *Acetobacterium* isolate (Frings et al. 1994). In both cases, TEA was degraded through DEA and MEA to end products such as ammonia, glycoaldehyde, and acetate.

A separate degradative pathway has been described for *Erwina caroto-vora*, which indicates growth using ethanolamine or monoisopropanolamine as the sole nitrogen source, resulting in the production of acetaldehyde or propionaldehyde, respectively (Fig. 2) (Jones et al. 1973). However, this microorganism was unable to use MEA or MIPA as a sole source of carbon for growth.

VIII. Aquatic Toxicology

The effects of the alkanolamines on aquatic organisms are of particular importance because these compounds would be expected to partition primarily to the water column. Toxicity is routinely assessed by addressing the potential adverse impact (death, impairment of growth or reproduction) to aquatic biota based upon acute or chronic exposure. Acute effects reflect an organism's response to short-term exposures, while chronic effects occur following repeated or long-term exposure.

The use of chronic and acute toxicity tests has gained widespread support as a mechanism for assessing potential risk of chemicals to aquatic biota. These tests are used to quantify the relative impact of a chemical during exposure to the organism. Acute toxicity tests have been used extensively to evaluate potential hazards to aquatic ecosystems with the endpoints of toxicity expressed as LC_{50} or EC_{50}. The LC_{50} or EC_{50} values are used as a measure of aquatic toxicity and are usually reported within a defined time period (24–96 hr). Toxicity endpoints are routinely expressed in milligrams per liter (mg/L), and the U.S. EPA (Zucker 1985) has ranked LC_{50} values according to the following descriptive categories:

< 0.1 mg/L, very highly toxic
$0.1–1.0$ mg/L, highly toxic
$> 1–10$ mg/L, moderately toxic
$> 10–100$ mg/L, slightly toxic
< 100 mg/L, practically nontoxic

Another common endpoint for toxicity tests is to report the concentration or dose that has minimal or nonexistent effect such as the no-observed-effect level (NOEL). The results from aquatic toxicity testing have been used to identify and evaluate potential adverse effects likely to occur to aquatic organisms and play an integral part in determining the overall risk of a chemical to the environment.

Numerous studies have been conducted on the aquatic toxicology of the alkanolamines. Some of these results are listed in Tables 10 through 14. This information includes a range of endpoints, not all of which are associated with a specific, recognized, toxicological category. In these cases, no attempt was made to classify the relative toxicity of compounds to organisms.

It is also important to note the arrangement of the information in Tables 10–14. The current phylogenetic classification system routinely used by taxonomists is based upon grouping living organisms into five distinct kingdoms: eubacteria, protists, fungi, animals, and plants. For purposes of this review, the tables were arranged to reflect this classification system, with the relevant aquatic toxicity for each compound divided into the corresponding five headings.

A wide range of single-species toxicity tests has been conducted on the alkanolamines. Most available data have been reported for either MEA, DEA, or TEA (Tables 10–12). These compounds have been extensively evaluated using a series of test organisms, including single-celled organisms, algae, invertebrates (e.g., *Daphnia*), and freshwater fish. In general, available data for the alkanolamines as a family of compounds suggest low toxicity of these materials to the majority of the species studied. For example, the effect that ethanolamines exert on growth of single-celled organisms has been an important parameter for assessing the overall impact of the ethanolamines on aquatic biota. Bringmann and co-workers (Bringmann and Kühn 1980; Bringmann et al. 1980) reported that MEA, DEA, and TEA exerted a decrease in population growth of $> 5\%$ on the single-celled *Entosiphon sulcatum* and *Chilomonas paramecium* at concentrations of 300, 160, and 56 mg/L and 733, 1137, and 1768 mg/L, respectively (Tables 10–12). Algae and cyanobacteria appear to be more sensitive to the ethanolamines, with significant population growth decreases noted in the range of approximately 1-10 mg/L for MEA, 3-20 mg/L for DEA, and 2-715 mg/L for TEA (Bringmann and Kühn 1976, 1978, 1980; Bringmann et al. 1980).

One of the most widely used organisms for single-species toxicity testing has been the water flea, *Daphnia magna*. This organism and *Ceriodaphnia dubia* are also commonly used for monitoring the overall water quality of lakes, streams, and effluents. In general, the ethanolamines exhibited minimal toxicity to these invertebrates. The LC_{50} values (Tables 10-12) for MEA, DEA, and TEA reported for *Daphnia magna* were 140, 55-306, and 1390 mg/L, respectively (Bringmann and Kühn 1977; Cowgill and Milazzo 1991; LeBlanc 1980). DEA was slightly more toxic to *Ceriodaphnia dubia*, with LC_{50} values ranging from 29 (Cowgill et al. 1985) to 160 mg/L (Cowgill and Milazzo 1991). Taken together, these results demonstrate that the ethanolamines can be classified as practically nontoxic to slightly toxic on the basis of acute tests with *Daphnia magna* and *Ceriodaphnia dubia*.

The potential toxicity of the ethanolamines has also been evaluated using a wide range of freshwater fish, including rainbow trout, fathead minnow, golden orfe, and bluegill. The ethanolamines exhibit a very low toxicity to these fish species. The LC_{50}s for MEA range from approximately 150 mg/L for rainbow trout (Johnson and Finley 1980) to 2100 mg/L for fathead minnow (Newsome et al. 1991). DEA also exhibited a very low toxicity to the fish, with LC_{50} values for fathead minnow ranging from >100 mg/L (Ewell et al. 1986) to 47,000 mg/L (Newsome et al. 1991). TEA had even less of an impact than MEA or DEA, with LC_{50}s ranging from approximately 1800 mg/L for fathead minnow (Newsome et al. 1991) to >10,000 mg/L for the golden orfe (Juhnke and Lüdeman 1978).

Although limited toxicity testing has been completed for the isopropanolamines and butanolamines, studies to date suggest low toxicity of these materials to the majority of the species studied. For example, the isopropanolamines exhibited a very low toxicity to fish (Table 13). The LC_{50}s for MIPA ranged from approximately 210 mg/L to >5000 mg/L for goldfish. This same species of fish was even less sensitive to DIPA, with LC_{50}s of 1100 to >5000 mg/L (Bridie et al. 1979b). Only one study evaluated the toxicity of TIPA; in this study, TIPA was force-fed to carp, indicating threshold mortality >950 mg/L (Loeb and Kelly 1963). Very little additional information was available regarding toxicity of the isopropanolamines to plants, eubacteria, protista, invertebrates, or other vertebrates.

The potential aquatic toxicity of butanolamines was examined using algae, *Daphnia*, and freshwater fish (e.g., rainbow trout), organisms representing three major trophic levels (Table 14). MBA, DBA, and TBA were also tested for their phytotoxicity to the freshwater green alga *Selenastrum capricornutum* Printz. The EC_{50}s were determined on the endpoint of algal growth with reported values of 17, 123, and 292 mg/L for MBA, DBA, and TBA, respectively. Kenega and Moolenar (1979) demonstrated that fish

Table 10. Aquatic toxicology of ethanolamine.

Species	Method	Endpoint	Value (mg/L)	Reference
Eubacteria				
Blue-green algae *Anacystis aeruginosa*	S,U	Population growth (toxicity threshold for cell multiplication inhibition test)	7.5	Bringmann and Kühn 1976
Microcystis (Diplocystis) aeruginosa	S,U	Population growth (decrease in cell multiplication), 8 d	2.1	Bringmann and Kühn 1978
Protista				
Flagellate euglenoid *Entosiphon sulcatum*	S,U	Population growth (>5% decrease in growth), 72 hr	300	Bringman and Kühn 1980
Cryptomonad *Chilomonas paramecium*	M	Population growth (5% decrease in cell count), 48 hr	733	Bringham et al. 1980
Green algae *Scenedesmus quadricauda*	S,U	Population growth (decrease in cell multiplication), 8 d	0.970	Bringham and Kühn 1978
	S,U	Population growth (toxicity threshold for cell multiplication inhibition test)	0.970	Bringman and Kühn 1976
	S,U	Population growth (3% decrease in extinction coefficient), 7 d	0.750	Bringman and Kühn 1980
Animalia (Invertebratae)				

Organism	Conditions	Endpoint	Value	Reference
Water flea	U	Mortality; acute toxicity begins here	100	Apostol 1975
Daphnia magna	U	Mortality; chronic toxicity begins here	<1.0	Apostol 1975
	S,U	LC_{50}, 24 hr	140	Bringmann and Kühn 1977
Animalia (Vertebrata)				
Mosquitofish	S,U	LC_{50}, 96 hr	337.5	Wolverton et al. 1970
Gambusia affinis				
Bluegill	S,U	LC_{50}, 96 hr	329.16	Wolverton et al. 1970
Lepomis macrochirus	S,U	LC_{50}, 96 hr	>300	Johnson and Finley 1980
Goldfish	S,M	LC_{50}, 96 hr	170	Bridie et al. 1979b
Carassius auratus	S,M, pH 10.1	LC_{50}, 96 hr	>5000	Bridie et al. 1979b
	S,M, pH 10.1	LC_{50}, 24 hr	190	Bridie et al. 1979b
Fathead minnow	F,M	LC_{50}, 96 hr	2100	Newsome et al. 1991
Pimephales promelas				
Silver or golden orfe	S,U	LC_{50}	224, 525	Juhnke and Lüdeman 1978
Leuciscus idus				
Rainbow trout	S,U	LC_{50}, 96 hr	150	Johnson and Finley 1980
Oncorhynchus mykiss				
Clawed toad	S,U	LC_{50}, 48 hr	220	deZwart and Sloof 1987
Xenopus laevis				

S, Static conditions; F, flow through; U, Unmeasured test chemical concentration, nominal valves given; M, measured test chemical concentration; Diet, test chemical force fed to test organism; LC_{50}, lethal concentration for 50% of the population; EC_{50}, toxic effects observed in 50% of the population; NOEL, No-observed-effects level; NOEC, no-observed-effects concentration.

Table 11. Aquatic toxicology of diethanolamine.

Species	Method	Endpoint	Value (mg/L)	Reference
Eubacteria				
Blue-green algae	S,U	Population growth (decrease in cell multiplication), 8 d	17	Bringmann and Kühn 1978
Microcystis (Diplocystis) aeruginosa				
Protista				
Marine diatom		EC$_{50}$: total cell count	548	Cowgill et al. 1989
Skeletonema costatum		EC$_{50}$: total cell volume	523	Cowgill et al. 1989
Flagellate euglenoid	S,U	Population growth (> 5% decrease in cell count), 72 hr	160	Bringmann and Kühn 1980
Entosiphon sulcatum				
Cryptomonad	U	Population growth (5% decrease in cell count), 48 hr	1137	Bringmann et al. 1980
Chilomonas paramecium				
Green algae	S,U	Population growth (3% decrease in extinction coefficient), 7d	4.4	Bringmann and Kühn 1980
Scenedesmus quadricauda				
	S,U	Population growth (decrease in cell multiplication), 8 d	10	Bringmann and Kühn 1978
Selenastrium capricornutium Printz		EC$_{50}$, 96 hr	3.3, 3.6	Dill et al. 1982

		Endpoint	Value	Reference
Animalia (Invertebratae)				
Common starfish	S,U	Population growth (toxicity threshold for cell multiplication inhibition test)	4	Bringmann and Kühn 1976
Asterias forbesi	S,U	Population growth (toxicity threshold for cell multiplication inhibition test)	10	Bringmann and Kühn 1976
Scud	S,U	LC_{50}, 96 hr	>100	Ewell et al. 1986
Gammarus fasciatus				
Aquatic sow bug	S,U	LC_{50}, 96 hr	>100	Ewell et al. 1986
Asellus intermedius				
Oligochaete	S,U	LC_{50}, 96 hr	>100	Ewell et al. 1986
Lumbriculus variegatus				
Flatworm	S,U	LC_{50}, 96 hr	>100	Ewell et al. 1986
Dugesia tigrina				
Ramshorn snail	S,U	LC_{50}, 96 hr	>100	Ewell et al. 1986
Helisoma trivolvis				
Brine shrimp	S,U	LC_{50}, 96 hr	2800	Price et al. 1974
Artemia salina				
Water flea	S,U	LC_{50}, 48 hr	55	LeBlanc 1980
Daphnia magna	S,U	Mortality: no-effect level	<24	LeBlanc 1980
	S,U	LC_{50}, 24 hr	180	Bringmann and Kühn 1977
	S,U, ~20 °C	LC_{50}, 48 hr (geometric mean)	116	Cowgill et al. 1985
	S,U, ~24 °C	LC_{50}, 48 hr (geometric mean)	109	Cowgill et al. 1985
	S,U	LC_{50}, 96 hr	>100	Ewell et al. 1986
Daphnia magna Strauss	S,U	LC_{50}, 48 hr	116	Gersich et al. 1985
Daphnia magna	S,U	3-brood test	3.5	Cowgill and Milazzo 1991
	S,U	LC_{50}, 48 hr	306	Cowgill and Milazzo 1991

(continued)

Table 11. (*Continued*)

Species	Method	Endpoint	Value (mg/L)	Reference
Ceriodaphnia dubia	S,U, ~20 °C	LC_{50}, 48 hr	78	Cowgill et al. 1985
	S,U, ~20 °C	LC_{50}, 48 hr	119	Cowgill et al. 1985
	S,U, ~20 °C	LC_{50}, 48 hr	104	Cowgill et al. 1985
	S,U, ~20 °C	LC_{50}, 48 hr (geometric mean)	99	Cowgill et al. 1985
	S,U, ~24 °C	LC_{50}, 48 hr (geometric mean)	30	Cowgill et al. 1985
	S,U,	3-brood test	19	Cowgill and Milazzo 1991
	S,U,	LC_{50}, 48 hr	160	Cowgill and Milazzo 1991
Animalia (vertebrata)				
Mosquitofish				
Gambusia affinis	S,U,	Median tolerance unit (TL_m), 96 hr	1400	Wallen et al. 1957
	S,U	Mortality (no effect at this level)	>320	Wallen et al. 1957
Bluegill	S,U	LC_{50}, 24 hr	1850	Turnbull et al. 1954
Lepomis macrochirus			2100	
Goldfish	S,M	LC_{50} pH 7.0, 24 hr	>5000	Bridie et al. 1979b
Carassius auratus	S,M	LC_{50} pH 9.7, 24 hr	80	Bridie et al. 1979b
Silver or golden orfe	S,U	LC_{50}	1430	Juhnke and Lüdeman 1978
Leuciscus idus			1850	
Fathead minnow	S,U	LC_{50} fry, 96 hr	1480	Mayes et al. 1983
Pimephales promelas				

Species		Endpoint	Value	Reference
	S,U	LC$_{50}$ juvenile, 96 hr	1550	Mayes et al. 1983
	S,U	LC$_{50}$ subadult, 96 hr	1370	Mayes et al. 1983
	S,U	LC$_{50}$, 96 hr (Geometric Mean)	1460	Mayes et al. 1983
	S,U	LC$_{50}$, 96 hr	>100	Ewell et al. 1986
	M,F	LC$_{50}$, 96 hr	47,000	Newsome et al. 1991
Clawed toad *Xenopus laevis*	S,U	LC$_{50}$, 48 hr	1174	deZwart and Sloof 1987
Sheepshead minnow *Cyprinodon variegatus*	S	LC$_{50}$, 96 hr	>540	Heitmuller et al. 1981
Plantae[a]				
Duckweed *Lemna gibba* G3	Plants	748[a]	216[a]	Cowgill et al. 1991
	Fronds	752	216	
	Dry Weight	1057	360	
Duckweed *Lemna minor* 6591	Plants	1525	600	
	Fronds	1504	600	
	Dry Weight	1514	600	
Duckweed *Lemna minor* 7101	Plants	1746	600	
	Fronds	2136	600	
	Dry Weight	2350	600	
Duckweed *Lemna minor* 7102	Plants	1035	360	
	Fronds	999	600	
	Dry Weight	1596	360	
Duckweed *Lemna minor* 7136	Plants	1355	216	
	Fronds	1314	600	
	Dry Weight	544	216	

[a]For plants, endpoint was 7-d phytotoxicity test, EC$_{50}$ m mg/L. Phytotoxicity of a compound was established by determining reduction in number of plants or fronds after exposure compared to control. In column 4, mg/L refers to 7-d phytotoxicity test NOEL.

Table 12. Aquatic toxicology of triethanolamine.

Species	Method	Endpoint	Value (mg/L)	Reference
Eubacteria				
Blue-green algae *Microcystis (Diplocystis) aeurginosa*	S,U	Population growth (decrease in cell multiplication), 8 d	715	Bringmann and Kühn 1978
Protista				
Cryptomonad *Chilomonas paramecium*	U	Population growth (5% decrease in cell count), 48 hr	1768	Bringmann et al. 1980
Flagellate euglenoid *Entosiphon sulcatum*	S,U	Population growth (>5% decrease in population growth), 72 hr	56	Bringmann and Kühn 1980
Green algae *Scenedesmus quadricauda*	S,U, Neutral pH	Population growth (toxicity threshold for cell multiplication inhibition test)	2	Bringmann and Kühn 1976
	S,U, Not neutral pH	Population growth (toxicity threshold for cell multiplication inhibition test)	715	Bringmann and Kühn 1976
	S,U	Population growth (3% decrease in extinction value), 7d	19	Bringmann and Kühn 1980
	S,U	Population growth (decrease in cell multiplication), 8d	715	Bringmann and Kühn 1978
Scenedesmus subspicatus		EC$_{50}$, 48 hr	470	Kühn and Pattard 1990
		EC$_{50}$, 48 hr	750	Kühn and Pattard 1990

Species		Test	Value	Reference
Animalia (Invertebrata)				
Water flea	S,U	LC_{50}, 24 hr	1390	Bringmann and Kühn 1977
Daphnia magna		EC_{50}, 24 hr	2038	Kühn et al. 1989
		NOEC; reproduction most sensitive parameter; parent animal mortality at nominal concentration; 21 d	16	Kühn et al. 1989
Brine shrimp	S,U	LC_{50}, 24 hr	5600	Price et al. 1974
Artemia salina				
Animalia (Vertebrata)				
Goldfish	S,M	LC_{50}, 24 hr	>5000	Bridie et al. 1979b
Carassius auratus				
Silver or golden orfe	S,U	LC_{50}	>10,000	Juhnke and Lüdemann 1978
Leuciscus idus	S,U	LC_{50}	>10,000	Juhnke and Lüdemann 1978
Aholehole	S,U	Behavior: no-effect level; 2 min	20	Hiatt et al. 1953
Kuhlia sandvicensis				
Common mirror-colored carp	Diet,U	Mortality: no-effect level; 45 hr	>176–213	Loeb and Kelly 1963
Cyprinus carpio				
Fathead minnow	F,M	LC_{50}, 96 hr	1800	Newsome et al. 1991
Pimephales promelas				

Table 13. Aquatic toxicology of isopropanolamine.

Vertebrate species	Method	Endpoint	Value (mg/L)	Reference
Monoisopropanolamine				
Goldfish	S,M, pH 7.0	LC$_{50}$, 24 hr	>5000	Bridie et al. 1979b
Carassius auratus	S,M, pH 9.9	LC$_{50}$, 24 hr	220	Bridie et al. 1979b
	S,M, pH 9.9	LC$_{50}$, 96 hr	210	Bridie et al. 1979b
Fathead minnow	F,M	LC$_{50}$, 96 hr	2500	Newsome et al. 1991
Pimephales promelas				
Clawed toad	S,U	LC$_{50}$, 48 hr	420	deZwart and Sloof 1987
Xenopus laevis				
Diisopropanolamine				
Goldfish	S,U	LC$_{50}$, 24 hr	>5000	Bridie et al. 1979b
Carassius auratus	S,U , pH 9.7	LC$_{50}$, 24 hr	1100	Bridie et al. 1979b
Clawed toad	S,U	LC$_{50}$, 48 hr	410	deZwart and Sloof 1987
Xenopus laevis				
Triisopropanolamine				
Carp	Diet,U	Mortality (no-effect level), 46 hr	>950	Loeb and Kelly 1963
Cyprinus carpio				

Table 14. Aquatic toxicology of mono-, di-, and triibutanolamines.

Species	Method	Endpoint	Value (mg/L)	Reference
Butanolamine				
Protista				
Green algae	S,U	EC$_{50}$ (growth), 120 hr	17	Brown et al. 1994
Selenastrium capricornutum Printz		NOEC	5	
Animalia (Invertebrata) Water flea	S,U	LC$_{50}$, 48 hr	85	Servinski et al. 1993
Daphnia magna Strauss				
Animalia (Vertebrata) Rainbow trout *Oncorhynchus mykiss* Walbaum	S,U	LC$_{50}$, 96 hr	102	Servinkski et al. 1993
Dibutanolamine				
Protista				
Green algae	S,U	EC$_{50}$ (growth), 120 hr	123	Milazzo et al. 1994
Selenastrium capricornutum Printz		NOEC	62.5	
Animalia (Invertebratae) Water flea	S,U	LC$_{50}$, 48 hr	228	Servinski et al. 1993
Daphnia magna Strauss				

(continued)

Table 14. (*Continued*)

Species	Method	Endpoint	Value (mg/L)	Reference
Animalia (Vertebrata) Rainbow trout *Oncorhynchus mykiss* Walbaum	S,U	LC$_{50}$, 96 hr	196	Servinski et al. 1993
Tributanolamine Protista Green algae *Selenastrium capricornutum* Printz	S,U	EC$_{50}$ (growth), 120 hr NOEC	292 125	Milazzo et al. 1994
Animalia (Invertebrata) Water flea *Daphnia magna* Strauss	S,U	LC$_{50}$, 48 hr	643	Servinski et al. 1993
Animalia (Vertebrata) Rainbow trout *Oncorhynchus mykiss* Walbaum	S,U	LC$_{50}$, 96 hr	625	Servinski et al. 1993

and *Daphnia* were appropriate indicators for toxicity of chemicals to aquatic life since aquatic animals were more sensitive indicators of toxic effects than plants or algae. The butanolamines could be considered practically nontoxic to slightly toxic, with LC_{50}s ranging from 85 to 643 mg/L for *Daphnia* and from 102 to 625 mg/L for freshwater fish. The lowest LC_{50} values were reported for MBA, with the relative toxicity to *Daphnia* and fish decreasing with increasing substitution of the butanolamine (e.g., MBA > DBA > TBA). These results demonstrate that the butanolamines are unlikely to elicit adverse environmental effects.

Summary

This review provides a summary of current information available on the environmental fate and aquatic toxicology of the alkanolamines. Because these materials are widely used, there is a need to understand their fate and effects in the environment. This assessment was confined to information regarding selected physical properties of the alkanolamines as well as their potential for degradation in the atmosphere, soil, surface water, and groundwater. In addition, their relevant aquatic toxicological information and bioconcentration potential were evaluated.

In general, the alkanolamines have high water solubilities and low to moderate vapor pressures. Some are solids whereas others are liquids at room temperature. Aqueous solutions of the alkanolamines are basic, with the pK_as decreasing with increased alkyl substitution. Predictions of the environmental distribution of these compounds, based on a unit world model of Mackay and Paterson, suggested that alkanolamines would partition primarily into the aqueous compartment at equilibrium, with the remainder distributed to the atmosphere. Only a very small fraction of these materials is expected to sorb to soil or sediments. However, adsorption mechanisms other than partitioning into the soil organic layer were not considered in this model. Since polar compounds may sorb to soil by alternate mechanisms, this model may underestimate the true adsorption potential and subsequent environmental distribution of the alkanolamines. Future work with these compounds should focus on other types of adsorption mechanisms that could impact the environmental distribution of the alkanolamines.

Although only small amount of the alkanolamines are expected to partition to the atmosphere, they are expected to be removed by reactions with photochemically generated hydroxyl radicals. They may also be removed from the atmosphere by precipitation, due to their high water solubility. Because of the relatively low levels expected to be present in the atmosphere and the relatively short half-lives, the alkanolamines are not expected to adversely impact air quality. Alkanolamines have also been shown to be highly susceptible to biodegradation and are not expected to persist in the environment. Results from numerous studies have shown that these materials undergo rapid biodegradation in soil, surface waters, and wastewater treatment plants. Degradation rates for these compounds may vary, with half-lives routinely in the range of 1 d to 2 wk, depending on the length of acclimation period and other environmental factors. The relatively low bioconcentration factor (BCF) values reported for the alkanolamines indicate that they would not be expected to bioconcentrate in aquatic organisms. Available data on the toxicity of the alkanolamines to aquatic organisms suggest low toxicity to the majority of the species studied.

Based on the facts that alkanolamines exhibit low aquatic toxicity, are shown to biodegrade in a wide range of environments, and exhibit no tendency to bioaccumulate, the routine manufacturing, use, and disposal of these materials are not expected to adversely impact the environment. With increased emphasis by consumers and regulatory agencies for industry to develop products that are "environmentally friendly," these properties of the alkanolamines make them an attractive choice for a wide range of applications.

Acknowledgments

We thank Carol Jones and Neil Hawkins for their support in preparation of this manuscript. We also extend our appreciation to Gary Klecka, Robert West, Monte Mayes, and Richard Brown for their technical review and valuable comments on the manuscript.

References

Alexander HC, Batchelder TL (1975) The pollution evaluation of compounds. The Dow Chemical Company, Midland, MI (unpublished report).

Alexander HC, McCarty WM, Bartlett EA, Syverud AN (1982) Aqueous odor and taste threshold values of industrial chemicals. Am Water Works Assoc J 74(11): 595–599.

Alexander M (1985) Biodegradation of organic chemicals. Environ Sci Technol 18: 106–111.

Anderson G (1979) Nitrosamines in cosmetics. J Cosmet Toilet 94:65–68.

Apostol S (1975) Ethanolamine toxicity in aquatic invertebrates (Engl. abstract). Stud Cercet Biol Ser Biol Anim 27(4):345–351.

Atkinson R (1988) Estimation of gas-phase hydroxyl radical rate constants for organic chemicals. Environ Toxicol Chem 7:435–442.

Ayanaba A, Alexander M (1974) Transformation of methylamines and formation of a hazardous product, dimethylnitrosamines, in samples of treated sewage and lake water. J Environ Qual 2:83–89.

Bailey RE, Brzak KA, Gonsior SJ, Harms DW, Hopkins DL, Piasecki DA, Reim RE, Voos-Esquivel CA, Warriner JP (1991) Diethylenetriamine: environmental fate in sewage, lake water, and soil. EPA 40-9139414 (OTS0531303). U.S. Environmental Protection Agency, Washington, DC.

Batchelder TL, Rhinehart WL (1977) BOD of TIPA using Midland City seed. The Dow Chemical Company, Midland, MI (unpublished report).

Birch RR, Fletcher RJ (1991) The application of dissolved inorganic carbon measurements to the study of aerobic biodegradability. Chemosphere 23(4):507–524.

Boethling RS, Alexander M (1979) Microbial degradation of organic compounds at trace levels. Environ Sci Technol 13(8):989–991.

Bridie AL, Wolff CJM, Winter M (1979a) BOD and COD of some petrochemicals. Water Res 13(7):627–630.

Bridie AL, Wolff, CJM, Winter M (1979b) The acute toxicity of some petrochemicals to goldfish. Water Res 13(7):623–626.

Bringmann G, Kühn R (1976) Limiting values for the damaging action of water pollutants to bacteria *(Pseudomonas putida)* and green algae *(Scenedesmus quadricauda)* in the cell multiplication inhibition test (Engl. summary). Z Wasser Abwasser Forsch 10(3–4):87–89.

Bringmann G, Kühn R (1977) Results of the damaging effect of water pollutants on *Daphnia magna.* Z Wasser Abwasser Forsch 10(5):161–166.

Bringmann G, Kühn R (1978) Testing of substances for their toxicity threshold: model organisms *Microcystis (diplocystis) aeruginosa and Scenedesmus quadricauda.* Mitt Int Ver Theor Angew Limnol 21:275–284.

Bringmann G, Kühn R (1980) Comparison of the toxicity thresholds of water pollutants to bacteria, algae, and protozoa in the cell multiplication inhibition test. Water Res 14(3):231–241.

Bringmann G, Kühn R, Winter A (1980) Determination of biological damage from water pollutants to protozoa. III. Saporozoic flagellates (Engl. abstract). Z Wasser Abwasser Forsch 13(5):170–173.

Brown RP, Milazzo DP, Servinski MF (1994) Monobutanolamine: the toxicity to the green alga *Selenastrum capricornutum* Printz. The Dow Chemical Company, Midland, MI (unpublished report).

Budavari S (ed) (1989) The Merck Index, 11th ed. Merck & Co., Inc., Rahway, NJ.

Chou WL, Speece RE, Siddiql RH (1979) Acclimation and degradation of petrochemical waste water components by methane fermentation. Biotechnol Bioeng Symp 8:391–414.

Cleveland CB, Ulmer JJ (1995) Aerobic soil metabolism study of triisopropanolamine (TIPA): a 2, 4-D requested moiety study. DowElanco, Indianapolis IN (unpublished report).

Cowgill UM, Takahashi IT, Applegath SL (1985) A comparison of the effect of four benchmark chemicals on *Daphnia magna* and *Ceriodaphnia dubia-affinis* tested at two different temperatures. Environ Toxicol Chem 4(35):415–422.

Cowgill UM, Milazzo DP, Landenberger BD (1989) Toxicity of nine benchmark chemicals to *Skeletonema costatum*, a marine diatom. Environ Toxicol Chem 8(5):451–455.

Cowgill UM, Milazzo DP (1991) The sensitivity of *Ceriodaphnia dubia* and *Daphnia magna* to seven chemicals utilizing the three-brood test. Arch Environ Contam Toxicol 20:211–217.

Cowgill UM, Milazzo DP, Landenberger BD (1991) The sensitivity of *Lemna gibba* G-3 and four clones of *Lemna minor* to eight common chemicals utilizing the 7-day test. J Water Pollut Control Fed 63:991–998.

Davis JW (1993) Physico-chemical factors influencing ethyleneamine sorption to soil. Environ Toxicol Chem 12:27–35.

deZwart D, Slooff W (1987) Toxicity of mixtures of heavy metals and petrochemicals to *Xenopus laevis*. Bull Environ Contam Toxicol 38(2):345-351.

Dill DC, Mayes MA, Shier QV (1982) The toxicity of chemicals to the freshwater green alga, *Selenastrum capricornutum* Printz. The Dow Chemical Company, Midland, MI (unpublished report).

Eisenreich B, Looney B, Thornton JD (1981) Airborne organic contaminants in the Great Lakes ecosystem. Environ Sci Technol 15(1):30-38.

Ewell WS, Gorsuch JW, Kringle RO, Robillard KA, Spiegel RC (1986) Simultaneous evaluation of the acute effects of chemicals on seven aquatic species. Environ Toxicol Chem 5(9):831-840.

Fattakhova AN, Ofitserov EN, Garusov AV (1991) Cytochrome P-450-dependent catabolism of triethanolamine in *Rhodotorula mucilaginosa*. Biodegradation 2: 107-113.

Frings J, Wondrak C, Schink B (1994) Fermentative degradation of triethanolamine by a homoacteogenic bacterium. Arch Microbiol 162:103-107.

Gannon JE, Adams MC, Bennett ED (1978) Microbial degradation of diethanolamine and related compounds. Microbios 23(91):7-18.

Gaudy AF (1972) Biochemical oxygen demand. In: Mitchell R (ed) Water Pollution Microbiology. Wiley-Interscience, New York, pp 305-332.

Gerike P, Fisher WK (1979) A correlation study of biodegradability determinations with various chemicals in various tests. Ecotoxicol Environ Saf 3:159-173.

Gersich FM, Blanchard FA, Applegath SL, Park CN (1985) The precision of daphnid (*Daphnia magna* Strauss, 1820) static acute toxicity tests. The Dow Chemical Company, Midland, MI (unpublished report).

Gerstl Z (1990) Estimation of organic chemical sorption by soils. J Contam Hydrol 6:357-375.

Gonsior SJ, West RJ (1991) Biodegradation of triethanolamine in soil and activated sludge. The Dow Chemical Company, Midland, MI (unpublished report).

Heitmuller PT, Hollister TA, Parrish PR (1981) Acute toxicity of 54 industrial chemicals to sheepshead minnows *Cyprinodon variegatus*. Bull Environ Contam Toxicol 27(5):596-604.

Hiatt RW, Naughton JJ, Matthews DC (1953) Effects of chemicals on a schooling fish, *Kuhlia sandvicensis*. Biol Bull 104:28-44.

Howard PH, Banerjee S (1984) Interpreting results from biodegradability tests of chemicals in water and soil. Environ Toxicol Chem 3:551-562.

Howard PH (ed) (1990) Handbook of Environmental Fate and Exposure Data for Organic Chemicals, Vol. II: Solvents. Lewis Publishers, Chelsea, MI.

Howard PH (1991) Atmospheric oxidation program. Version 1.3. Syracuse Research Corp., Syracuse, NY.

Huntziger O, Von Letyoeld LH, Zoeteman BC (1978) Aquatic pollutants: transformation and biological effects. Pergamon Press, New York.

Johnson WW, Finley MT (1980) Handbook of acute toxicity of chemicals to fish and aquatic invertebrates. Resource Publ 137. U.S. Fish and Wildlife Service, U.S. Department of Interior, Washington, DC.

Jones A, Turner JM (1973) Microbial metabolism of amino alcohols; 1-aminopropan-2-ol and ethanolamine metabolism via propionaldehyde and acetaldehyde in a species of *Pseudomonas*. Biochem J 134:167-182.

Jones A, Faulkner A, Turner JM (1973) Microbial metabolism of amino alcohols;

metabolism of ethanolamine and 1-aminopropan-2-ol in species of *Erwinia* and the roles of amino alcohol kinase and amino alcohol *s*-phosphate phosphorylase in aldehyde formation. Biochem J 134:959–968.

Juhnke I, Lüdeman D (1978) Results of the investigation of 200 chemical compounds for acute fish toxicity with the golden orfe test (Engl. abstract). Z Wasser Abwasser Forsch 11(5):161–164.

Kenaga EE, Moolenar RJ (1979) Fish and *Daphnia* toxicity as surrogates for aquatic vascular plants and algae. Environ Sci Technol 13:1479–1480.

Klecka GM (1985) Biodegradation. In: Neely WB, Blau GE (eds) Environmental Exposure from Chemicals, Vol. 1. CRC Press, Boca Raton, FL, pp 109–155.

Klecka GM, Landi LP (1985) Evaluation of the OECD activated sludge, respiration inhibition test. Chemosphere 14(9):1239–1251.

Krieger MS (1995) Aerobic aquatic metabolism of ^{14}C-triisopropanolamine (TIPA). DowElanco, Indianapolis, IN (unpublished report).

Kuenemann P, DeMorsier A, Vasseur P (1992) Interest of carbon-balance in ready biodegradability testing. Chemosphere 24(1):63–69.

Kühn R, Pattard M, Pernak K, Winter A (1989) Results of the harmful effects of water pollutants to *Daphnia magna* in the 21 day reproduction test. Water Res 23(4):501–510.

Kühn R, Pattard M (1990) Results of the harmful effects of water pollutants to green algae *(Scenedesmus subspicatus)* in the cell multiplication inhibition test. Water Res 24(1):31–38.

Lamb CB, Jenkins GF (1952) BOD of synthetic organic chemicals. In: Bloodgood DE (ed) Proceedings of the 8th Indiana Waste Conference, Engineering Bulletin. Purdue University, West Lafayette, IN, pp 326–339.

LeBlanc GA (1980) Acute toxicity of priority pollutants to water flea *(Daphnia magna)*. Bull Environ Contam Toxicol 24(5):684–691.

Lewis GE (1992) Determination of pH of aqueous alkanolamines. The Dow Chemical Company, Midland, MI (unpublished report).

Loeb HA, Kelly WH (1963) Acute oral toxicity of 1,496 chemicals force fed to carp. Special Scientific Report — Fisheries, No. 471. U.S. Fish and Wildlife Service, Washington, DC.

Long, MW Jr (1955) Physical properties of various alkanolamines. The Dow Chemical Company, Midland, MI (unpublished report).

Lyman WJ, Reehl WF, Rosenblatt DH (1982) Handbook of Chemical Property Estimation Methods. McGraw-Hill, New York.

Mackay D, Paterson S (1981) Calculating fugacity. Environ Sci Technol 15:1006–1014.

Mayes MA, Alexander HC, Dill DC (1983) A study to assess the influence of age on the response of fathead minnows in static acute toxicity tests. Bull Environ Contam Toxicol 31(2):139–147.

Medicinal Chemistry Project (1989) Med. Chem. release 3.54. Daylight Chemical Information Systems, Inc., Irvine, CA.

Milazzo DP, Servinski MF, Martin MD (1994) Monobutanolamine, dibutanolamine, tributanolamine: the toxicity to the green alga *Selenastrum capricornutum* Printz. The Dow Chemical Company, Midland, MI (unpublished report).

Mills AL, Alexander M (1976) *N*-nitrosamine formation by cultures of several microorganisms. Appl Environ Microbiol 31(6):892–895.

Mills EJ, Stack VT (1952) Biological oxidation of synthetic organic chemicals. In: Proceedings of the 8th Industrial Waste Conference. Eng Bull Purdue Univ Ext Ser 83:492–517.

Mills EJ, Stack VT (1954) Acclimation of microorganisms for the oxidation of pure organic chemicals. In: Proceedings of the 9th Industrial Waste Conference. Eng Bull Purdue Univ Ext Ser 9:449–464.

Mirvish SS (1975) Formation of *N*-nitroso compounds: chemistry, kinetics, and *in vivo* occurrence. Toxicol Appl Pharmacol 31:325–351.

Muller M, Klein W (1991) Estimating atmospheric degradation processes by SARs. Sci Total Environ 109/110:261–273.

Newsome LD, Johnson DE, Upnick RL, Broderius SJ, Russom CL (1991) A QSAR study of the toxicity of amines to the fathead minnow. Sci Total Environ 109/110:537–552.

Organization for Economic Cooperation and Development (OECD) (1993) OECD guidelines for testing of chemicals. OECD, Paris, France.

Pitter P (1976) Determination of biological degradability of organic substances. Water Res 10:231–235.

Price KS, Wagy GT, Conway RA (1974) Brine shrimp bioassay and seawater BOD of petrochemicals. J Water Pollut Control Fed 46(1):63–77.

Richardson ML, Webb KS, Gouch TA (1979) The detection of some *n*-nitrosamines in the water cycle. Ecotoxicol Environ Saf 4:207–212.

Sax NI, Lewis RJ (1987) Hawley's Condensed Chemical Dictionary, 11th ed. Van Nostrand Reinhold, New York.

Servinski MF, Richardson CH, Brown RP (1993) Monobutanolamine, dibutanolamine, tributanolamine: static acute toxicity to the water flea, *Daphnia magna* Strauss and the rainbow trout, *Oncorhynchus mykiss* Walbaum. The Dow Chemical Company, Midland, MI (unpublished report).

Struijis JM, Stoltenkamp-Wouterse J, Dekkers ALM (1995) A rationale for the appropriate amount of inoculum in ready biodegradability tests. Biodegradation 6:319–327.

Sugatt RH, O'Grady DP, Banerjee S, Howard PH, Gledhill WE (1984) Shake flask biodegradation of 14 commercial phthalate esters. Appl Environ Microbiol 47:601–606.

The Dow Chemical Company (1988) Physical properties of the alkanolamines. Form no. 111-1227-88. The Dow Chemical Company, Midland, MI.

Thomann RV (1989) Bioaccumulation model of organic chemical distribution in aquatic food chains. Environ Sci Technol 23:699–707.

Turnbull H, DeMann JG, Weston RF (1954) Toxicty of various refinery materials to fresh water fish. Ind Eng Chem 46(2):324–333.

Urano K, Kato Z (1986) A method to classify biodegradabilities of organic compounds. J Hazard Mater. 13:135–145.

Wallen LE, Greer WC, Lasater R (1957) Toxicity to *Gambusia affinis* of certain pure chemicals in turbid waters. Sewage Ind Wastes 29(6):695–711.

West RJ (1995) The biodegradation of diisopropanolamine. The Dow Chemical Company, Midland, MI (unpublished report).

West RJ, Gonsior SJ (1996) Biodegradation of triethanolamine. Environ Toxicol Chem 15(4):472–480.

Williams GR, Calley AG (1981) The biodegradation of diethanolamine and triethanolamine by a yellow gram-negative rod. J Gen Microbiol 128:1203–1209.

Wolverton BC, Harrison DC, Voigt RC (1970) Toxicity of decontamination products. Tech rep AFATL-TR-70-68, (NTIS AD-879 811). U.S. Air Force Armament Laboratory, Eglin Air Force Base, FL.

Yordy JR, Alexander M (1980) Microbial metabolism of *n*-nitrosodiethanolamine in lake water and sewage. Appl Environ Microbiol 39(3):559–565.

Yordy JR, Alexander M (1981) Formation of *N*-nitrosodiethanolamine from diethanolamine in lake water and sewage. J Environ Qual 10(3):266–270.

Young RHF, Ryckman DW, Buzzell JC Jr (1968) An improved tool for measuring biodegradability. J Water Pollut Control Fed 40(8):354–368.

Zahn R, Wellens H (1980) Examination of biological degradability through the batch method — further experience and new possibilities of usage (Engl. abstract). Z Wasser Abwasser Forsch 13(1):1–7.

Zucker E (1985) Acute toxicity test for freshwater fish. (EPA-540/9-85-006). U.S. Environmental Protection Agency, Washington, DC.

Manuscript received February 27, 1996; accepted May 9, 1996.

Index